WINNING THE FUTURE
From Bhakra to Narmada, Tehri, Rajasthan Canal

WINNING THE FUTURE

From Bhakra to Narmada, Tehri, Rajasthan Canal

B.G. VERGHESE

under the auspices of
CENTRE FOR POLICY RESEARCH, New Delhi

KONARK PUBLISHERS PVT LTD

KONARK PUBLISHERS PVT LTD
A-149, Main Vikas Marg, Delhi 110092

Typeset by Dexter Computer Services, Delhi and printed at
Ravindra Offset Press, Delhi 110032 (India)

Foreword

THE Centre for Policy Research has sought to focus on a variety of urgent issues that merit attention. The emerging water crisis; the environmental impacts of development; alternative courses of action or inaction and their outcomes in terms of poverty alleviation, equity and a better quality of life for the people, including India's large and growing tribal population. These are all appropriately on the national agenda. In any complex situation, nothing is exclusively black or white. Yet a balance must be struck and a reasonable via media found without undue delay. Time is not on India's side.

In this volume, Prof B.G. Verghese sets out to look at and answer critiques of four major water resource projects, Narmada, Tehri, the Indira Gandhi Nahar (Rajasthan Canal), each at various stages of construction, and a mature project like Bhakra-Pong where both costs and benefits are visible and measurable and no longer a matter of speculation. Bhakra and Pong show that large dams are not necessarily disasters.

While making a broad-brush presentation for the lay reader, Verghese has found it necessary to include a fair amount of detail in order to answer the sweeping and persistent criticism directed against these large projects. Reasoned justification has been provided in the face of emotional outbursts against them. The report of the Independent Review Mission on the Sardar Sarovar Project which he finds flawed and pretentious, has been comprehensively rebutted.

Current environmental controversies over a number of large projects and water resource developments in particular have tended to question the nation's present development path. Development has to be sustainable; but whether sustainability with rapidly growing numbers living on the margin of subsistence is possible without adequate development is a moot point. The social and political consequences of not being able to stay ahead of Malthus in this volatile day and age cannot be kept out of the reckoning.

Likewise, while the human rights of every individual and community are supremely important, the right to life of many millions cannot be ignored either and, weighed in the balance, must inevitably count for more. These two interests are, however, not incompatible. They can be harmonised, perhaps not always here and now, but certainly over time and space.

Nations and peoples must learn from experience. But there can be no shrinking from progress or a return to arcadia. Development cannot be an exercise in extreme pessimism which carries with it high political and economic costs.

The message in this volume is that India must, with prudence and confidence, seize the opportunity

Director
Centre for Policy Research,
New Delhi

V.A. PAI PANANDIKER

Acknowledgements

THIS work owes a great deal to a very large number of persons connected with the projects concerned and their critics, all of whom gave generously of their time to discuss the projects in their larger setting and explain their points of view.

Without attempting an exhaustive list, my grateful thanks are due to successive Secretaries to Government in the Ministry of Water Resources, the Central Water Commission, the Department of Power and the Ministry of Environment and Forests.

Officials of the World Bank in New Delhi and Washington.

The Chairman and officials of the Bhakra-Beas Management Board at Chandigarh, Talwara and Bhakra; officials of the Governments of Punjab and Haryana.

The Chairman and officials of the Indira Gandhi Nahar Pariyojana at Jaipur, Jaisalmer, Bikaner and along the Canal; the IGNP Command Area Development Authority and Colonisation Commissioner, Bikaner; and officials of the Rajasthan Government. The Indus Commissioner.

The Chairman and officials of the Tehri Hydro Development Corporation in Delhi, Rishikesh, Tehri and Dehra Dun.

Senior faculty at Roorkee University, especially in the Departments of Earthquake Engineering and Earth Sciences.

Dr Jai Krishna; Dr Vinod Gaur; B.B. Raj; R.B. Shah.

Members of the Narmada Control Authority, Indore.

The Chairman and officials of the Sardar Sarovar Nahar Nigam Ltd. at Gandhinagar, Baroda and Kevadia; officials of the Gujarat Government; the Vice-Chairman and Members of the Narmada Planning Group, Ahmedabad and Gandhinagar.

The Chairman and officials of the Narmada Valley Development Authority at Bhopal, Narmada Sagar, Maheshwar, Omkareshwar, Tawa and elsewhere in the NSP command; and officials of the M.P. Government.

Dr Anil Patel and his colleagues in ARCH-Vahini, Mangrol;

Medha Patkar of NBA; Baba Amte; Sunderlal Bahuguna; Sanjoy Ghose of URMUL, Bikaner; Nalini Jayal of INTACH, New Delhi; Sunil Roy; Vasudha Dhagamwar of MARG; New Delhi; and many other environmental activist groups.

At the Centre for Policy Research, I received every encouragement from the Director, Dr V.A. Pai Panandiker, and all assistance from the Librarian, Mr Kamaljit Kumar and his staff. Word-processing and revisions were diligently undertaken by Ms Neelam Grover and her colleagues.

B.G. VERGHESE

Conversion Table

Water flows and volumes can be expressed in foot-pound or metric measures. Though India has switched to the metric system, both measures are used in this volume, depending on the original source and literature available.

The following are foot-pound equivalents of metric measures and vice versa:

1 inch is 25.4 millimetres (mm) or 2.54 cm.

1 yard is 0.9144 metres; 1 metre is 1.094 yards.

1 mile is 1.6093 kilometres.

1 hectare is 2.47 acres.

100 hectares make one square kilometre.

1 cubic metre equals 35.315 cubic feet.

10,000 cubic metres makes one hectare-metre.

One million hectare-metres equals one cubic kilometre.

1233.48 cubic metres equals one million acre-feet.

One hectare-metre equals 8.107 acre-feet.

One cumec equals 35.315 cusecs.

One litre is equivalent to 0.22 gallons.

Glossary/Abbreviations

BBMB	Bhakra-Beas Management Board
B-C Ratio	Benefit-cost ratio
Bgl	Below ground level
CAD (A)	Command Area Development (Authority)
CAP	Canal Affected Person
Catchment	Drainage area of a river
CAT	Catchment Area Treatment
Command	Area commanded by an irrigation project
Cu m	Cubic metres
Cumec	Cubic metres per second
Cusec	Cubic feet per second
CWC	Central Water Commission
DVC	Damodar Valley Corporation
EPGA	Effective Peak Ground Acceleration
FRL	Full reservoir level
FSL	Full supply level
GOG	Government of Gujarat
GOI	Government of India
GOM	Government of Maharashtra
GOMP	Government of Madhya Pradesh
Ha	Hectare
Ha m	Hectare-metre
HFL	Highest flood level
IGNP	Indira Gandhi Nahar Pariyojana
INTACH	Indian National Trust for Art & Cultural Heritage

IRM	Independent Review Mission (of World Bank)
ISP	Indira Sagar Project
Kw	Kilowatt
Kwh	Kilowatt-hour (unit)
LA Act	Land Acquisition Act
Lpcd	Litres per capita day
m	Metres
m	Million
M	Magnitude of earthquake on Richter scale
MAF	Million acre-feet
MCE	Maximum Credible Earthquake
MDDL	Minimum Drawdown Level
M & E	Monitoring and Evaluation
MOEF	Ministry of Environment & Forests
MM	Intensity of earthquake on Modified Mercalli scale
MW	Megawatt
NBA	Narmada Bachao Andolan
NCA	Narmada Control Authority
NDS	Narmada Dharangast Samiti
NPG	Narmada Planning Group
NSP	Narmada Sagar Project
NWDT	Narmada Water Disputes Tribunal
O & M	Operation and maintenance
PGA	Peak ground acceleration
PMF	Probable maximum flood
R & R	Resettlement and Rehabilitation
SCARP	Salinity Control and Reclamation Project (Pakistan)
SSNNL	Sardar Sarovar Nahar Nigam Ltd
SSP	Sardar Sarovar Project
SYL	Sutlej-Yamuna Link

THDC	Tehri Hydro-Development Corporation
TVA	Tennessee Valley Authority
2/3-D analysis	Two-Three dimensional analysis
URMUL	Uttar Rajasthan Milk Union Ltd., Bikaner

Contents

1

Lost Horizon

RECENT years have seen the growth of an anti-dam movement worldwide, currently epitomised by the extraordinary hate campaign launched against the Sardar Sarovar Project. Twentyfive years ago, the Aswan High Dam was reckoned an environmental disaster by many critics. Today, because of it, Egypt is twice blessed by the Nile.

Controversies over large dams relate to a variety of issues: displacement, safety, and environmental losses among others. The benefits are more often than not glossed over. International groups have joined the fray. Motives have been impugned and not just these projects but the decision-makers, engineers and administrators responsible have also come to be seen as evil, anti-people. Discussion has become so emotionally charged as to render reasoned debate difficult and sometimes impossible.

Underlying much of the argument is a rejection of the current development path as being mindlessly growth-oriented, consumerist, globalised and exploitative. Large dams are seen as visible symbols of this trend. Greens, social activists, and alternative-society counter-culture ideologues have gathered at the barricades to defend the future as they see it. As a generalised theory this is a mistaken view and surely exaggerated in the Indian or Third World context where far too many are still graduating from survival to poverty. Attainment of the minimum decencies of life remains a distant goal for large sections of the population, with tribal communities most often at the bottom of the heap.

To quench a thirst, grow more food, generate clean energy and create employment and opportunity for millions is far from a consumerist catastrophe.

Facing Water Stress

All civilisations have grown along rivers or other sources of water, for water is the giver of life. Thus the great Indus and Ganges civilisations. Long before that, the inspired paintings of Narmada Man in the rock shelters of Bhimbetka, south of Bhopal, depicted warriors, hunters, a variety of animals — and a dam. So it was, too, along the Nile, the Tigris and Euphrates, the Yangtze, everywhere, as nomadism yielded to settlement.

Science has increasingly opened new vistas and found substitutes for many natural products or alternative sources of energy. But there is as yet no substitute for water which has over the ages been conserved and stored and spread over time and space through irrigation, aqueducts and pipelines to grow food and sustain cities and rural communities.

The earth is full of water, the oceans and polar ice accounting for so much of our planet. But only a small part of this occurs as usable rain, snow or ice that feed rivers, lakes or groundwater and not all of this is well distributed. Meanwhile, the demands on freshwater have continually increased with development and, especially, burgeoning population growth. On the other hand, the good earth has turned to desert through improper management of land and water resources and there is ample evidence of this in history the world over. So environmental concerns are obviously important and "sustainability" a critical pathway.

Large parts of the world are, or are fast becoming, water stressed, including much of northwestern, north central, western and southern India. This shortage is aggravating environmental degradation and human distress in a vicious cycle of poverty, unemployment and migration. The problems of global warming have added another cause for concern. It is against this background that the nation's efforts to conserve and harness its water resources must be seen.

Towards Sustainability

It should be no surprise that water resource development was a focal point of India's early Five Year Plans with projects like Bhakra, the DVC, Hirakud, Nagarjunasagar, Tungabhadra and many more to provide irrigation, flood control and hydro power. Jawaharlal Nehru aptly described them as Temples of Modern

India. There may have been errors of conception or cost-benefit analysis and faulty implementation in some of these developments. The country was lower on the learning curve. Understanding of the environmental aspects of development, though inherent in Gandhian thinking, was limited the world over and did not impact on development policies and programmes until the World Environment Conference in Stockholm in 1972. That same period saw the rise of the Chipko movement and a steady growth in environmental consciousness. The notion of sustainability followed; and now concerns about greenhouse gases.

These are legitimate concerns. But the basic environmental issues confronting the industrially advanced nations and developing societies differ. The major environmental problems of the North stem from affluence and waste; those of the South from the struggle for subsistence. Indira Gandhi was not wrong to tell the Stockholm Conference that poverty is the worst polluter. And, with growing populations, the pressure becomes greater.

Therefore, to look at Third World development through Western eyes and norms is misconceived. The advanced industrial nations of North America, Europe, Japan and Australia have much of their water resource development (incorporating many large dams) behind them, not ahead of them as do most developing nations. Nor do the temperate regions of the North by and large suffer the same rigours of aridity and crop-water management as those in the South. Is the developing world to forgo the options that the industrial nations exploited to the full and can it afford to do so? The nexus between the anti-dam movement in India and their Western counterparts would appear to rest in large part on a huge historical and agro-climatic fallacy.

The natural resource base of any country, its land, water and forests must be productively used and conserved, not merely preserved. This provides the foundation for agriculture, industry and urbanisation. Water resource development must aim at optimal redistribution of the monsoon. There are many ways of achieving this and no exclusive path. Hence the futility of argument about small versus large storages and surface versus ground water. This is altogether a wrong approach. With a highly skewed seasonal and spatial distribution of water, storages and transfers are required. Water has to get to people or they will move out in droves and migrate to greener pastures whenever the rains fail, for water not only supports life but sustains the productive process. Inter-basin transfers constitute one means of evening out spatial variations and India's National Water Grid, elements of which are gradually beginning to come into place, represent a certain long term insurance and regional water equity.

Interference with nature is not an abomination but must be done with great care and understanding. History is replete with tragic examples of nature's revenge where brashness rather than sensitivity has informed management of critical life support systems. But nature itself can be wayward and changing and man has had to adapt or compensate for these changes to survive.

India has no option but to harness its water resources to the extent possible or face increasing distress and destabilisation. Water quality too is of importance. Growing numbers and expanding economic activity is putting increasing pressure on limited resources. Careless management of effluents, return flows and emissions can abuse nature's regenerative capability beyond repair. Hence the need for measuring and mitigating adverse environmental impacts, including pollution.

It is within this frame that development must progress, with sustainability a critical criterion. But poorer societies cannot turn away from development either. That is not politically or socially sustainable, and poverty is no friend of the environment. Development goals and strategies may be debated and questioned but there is a minimum threshold that must be crossed to ensure all citizens the basic decencies of life and enjoyment of health and leisure. To shrink from that in search of a lost world of the past offers no solution.

Mired in Myth

The headwaters of India's rivers lie in its hills and forests. So it is inevitable that these areas will remain a major focus of water resource development. Ninety per cent of the country's mineral wealth also lies buried along a broad band of middle India. This triple convergence or coincidence of water, forests and minerals will compel increasing intrusion into what is also by and large a tribal belt. Exploitation of these resources must ensure development of the region and all of the people living there.

A number of myths have been propagated about the pristine nature of India's forests and forest-flood relationships. Silent Valley was said to be a unique rain-forest with a closed, ecosystem that had climaxed in conditions free of any external interference whatsoever. It was therefore seen as a scientific treasure-house with exotic endemic species which should not be touched, let alone overrun. To describe any good or lush stand of forest as a "rain-forest" or a Silent Valley, as has now become the fashion in some quarters, is absurd. Few people realise that, barring limited exceptions, very large parts of India's forests have been worked, felled, burnt and regenerated over the centuries. This would also be true of much of Nepal's forests and is true of the European and Japanese Alps and the Appalachian mountains as timber has always been in demand for housing, smelting, shipbuilding and warlike purposes and as industrial fuel. Conservation is a relatively new concept. There is also confusion about the ecological role of forests as against their productive role which vitiates discussion about compensatory afforestation and the upgradation of degraded forests and plantations. The productivity of Indian forests is rather low and there is no reason to believe that efforts to upgrade it are necessarily invasive or consumerist.

There are also a number of myths idealising the tribal condition. Why should not tribal people, like any other, be given the option to change that goes with the right to development. They must not be treated as wards by those who arrogate the right to determine that they should remain as they are. Their identity and culture, as of all other groups in India's vastly diverse plural society, must be protected. But this cannot mean that they be treated as anything less than full and equal partners in building a better society free of want, ignorance and disease. India represents one of the oldest living civilisations and has the largest tribal population in the world — close on 70 million. Efforts to categorise tribal populations as "indigenous people" in the Western idiom is doing violence to history and an attempt, howsoever unintended, to create divisions where none should exist.

Canada and Australia are "returning" native lands or land rights to the Inuit (Eskimos) and aborigines which the outsider forcibly usurped a century or more ago. A section of Australians, especially mining interests, is greatly concerned over a High Court ruling recognising native land title for Australian aborigines on a petition by a tribal leader, Eddie Mabo. The judgement overturned the long-held doctrine of terra nullius which held Australia to have been uninhabited when colonised by the white man. The judgement is, however, limited to vacant crown lands to which aboriginals can prove continuous traditional association. (Seth, July 1993) Likewise, Canada is constitutionally moving towards grant of home rule to its Indians, starting with the Eskimos or Inuit people in a new 850,000 square mile territory of Nunavut in the Arctic region inhabited by 22,000 Inuits. (Langan, June 1992)

These are steps that return territorial rights taken away by the white man from indigenous people. This is not the Indian tribal condition. The tribal people of India share a common traditional homeland with the rest of the population and are partners in a common enterprise. If it is to be acceptable, however, development cannot result in grief and little or no gain to those who are displaced and asked to make a national sacrifice at the cost of their hearths, fields and cultural identity for the creation of exclusive enclaves and a stream of benefits that go to other regions and other people. They too must be co-partners. Development at whose cost? And for whose benefit? These questions cannot be evaded.

It is undoubtedly true that in the past, ignorance, lack of awareness, arrogance and callousness may have resulted in neglect of and even extreme distress to those displaced by development. They were paid some simple compensation and thereafter perhaps left out of the reckoning. Compensation is not necessarily equivalent to rehabilitation and indeed, until fairly recently, that term was scarcely used or properly understood, administrative action often being limited to monetary compensation under the Land Acquisition Act. That is no longer possible. Nor was there an adequate understanding of environmental impacts, or policies in place to ensure prior action to avoid or mitigate adverse repercussions as part of the basic project design. Both these aspects are now central to all planning and development. The country and the world have both learnt from experience and therefore what happened yesterday, if tragic, does not necessarily have to be repeated tomorrow. Nor does prudence lie in being yesterday's men, forgoing the future.

The promise of development has not been anywhere near fully realised. There have been many gaps and disappointments. The reasons for this are many, but the disenchantment to which it has given rise has also resulted in questioning political structures, the development path, national goals and social relationships. Despair, cynicism, uncertainty and even anger have been manifest in a variety of pulls and pressures, the emergence of a new radicalism and a new democracy with the social and political awakening of the large sub-stratum of the demos that was long silent and inert.

Rejectionist Theories

In the churning that has followed, rejectionist theories too have been propagated, some with anti-development and anti-science overtones. There is a yearning for the past and calls for a return to nature and development on a human scale rather than through monolitic structures and behemoths. Hence the urge for small dams and little systems. There is no virtue in giganticism and it was Wittfogel's thesis that oriental despotisms grew out of large, centralised hydraulic-based civilisations. Nonetheless, small is not always beautiful any more than big is necessarily bad. Each has its place and none need exclude the other or a whole range in between or beyond. And many small component units can be organised into very large systems as much as a huge project can be operated and managed on a decentralised basis and in discrete segments.

As many of the easier and more obvious options of water resource development have been exhausted, resort has now to be taken to more difficult and complex possibilities. Many of these projects have come under severe criticism on a variety of counts — size, viability, displacement, safety, impacts, cost-benefit sharing. The Narmada Valley projects, the Tehri Dam, and the Rajasthan Canal (now known as the Indira Gandhi Nahar) have been in the eye of the storm and have attracted international attention. "Stop the dam" has become a revolutionary cry as evocative as "to the barricades" and "they shall not pass". Some of the premises and perceptions of the underlying environmental rhetoric are questionable or simplistic and smack of eco-fundamentalism. This evokes an equally irrational antagonism that could be as unfortunate. Real environmental concerns and real development opportunities for those most in need could be the unintended victim of excessive heat on a false agenda.

There is also a premise that Narmada, Tehri, the I.G. Nahar and similar developments do not help "people" but are products of a conspiracy between greedy technocracts, wily bureaucrats and corrupt politicians in cahoots with degenerate industrialists to rob the poor. This "elites" versus "people" syndrome keeps intruding into the debate. The dispossessed are "people", but anyone who has earned modest success or done well is "unpeopled" and placed among the ranks of the exploiting classes! This new demonology would appear almost to penalise success and glorify failure and poverty as virtues in themselves. Yes, the meek shall hopefully inherit the earth: but maybe God could do with a helping hand.

There is also in the new radicalism abroad a hint of the State being the real or ultimate enemy which must be overcome by the "people". This anarchist note is worrying. There is much reason to seek democratic reform and changes in political structures that ensure greater participation and accountability, and to "empower" people. The "people" are being invoked to struggle against "anti-people" projects like Narmada, Tehri and the I.G. Nahar that symbolise the "state", "elites" and a development path that is held to be wicked. Such definitional logic rests on dubious assumptions and is a prescription for a new kind of permanent revolution.

A section of international opinion has latched on these issues. Narmada has become a *cause celebre*.

If all large water resource projects are bad or are predictably going to sour within 30 years, as suggested by some critics, what about Bhakra-Pong? Has that not been transforming, as visible evidence suggests, or is it an emerging disaster and the fountainhead of terrorism in Punjab as some have painted it? The argument will sound bizarre to most, but it is being seriously propagated through platform, print and film.

It is against this larger background that this volume attempts to examine these four projects and answer the virulent criticism of them by environmental activists. Much of this hate campaign would appear to rest on an incomplete understanding of the facts and on a number of basic misconceptions. What is perhaps more surprising is that a large number of eminent and rational persons in diverse fields should echo these sentiments and periodically call for a review of these projects. This can only be attributed to a wider disenchantment with the country's development path and progress, loss of credibility in the system of governance, and development fatigue. Much of this is irrational. It is not based on a careful study of the facts and is emotional and visceral rather than cerebral. This shows how far the state has lost the power to communicate with its own people, a tragic and monumental failure. There has been a failure of the media too, with froth being confused with substance.

The Indira Gandhi Canal is still only partly complete but has begun to transform the Rajasthan desert. The Sardar Sarovar, Narmada Sagar and Tehri projects are at earlier stages of development and hold out large promise. The Narmada Valley projects will change the face of middle India. Tehri marks the first major step towards harnessing the immense resources of the Ganga-Brahmaputra system through Himalayan storages. Bhakra and Pong have turned Punjab and Haryana into India's bread basket, ending the PL 480 syndrome, a factor of enormous economic and political significance.

Together they represent a vision of the future. The early completion of these on-going projects will undoubtedly trigger positive change that could in some measure restore faith.

2

Minus Bhakra?

INDIA's first five year Plan (1952-57) essentially consisted of three major river valley projects, Bhakra-Nangal, Hirakud and the DVC, the Sindri Fertiliser and Chittaranjan Locomotive plants, and the community development programme. A decade later, when the Bhakra Dam was commissioned, Jawaharlal Nehru was to describe it with pride as a Temple of Modern India. Many derided this "white elephant" and wondered where the power would go and even whether the irrigation potential would be easily absorbed. In tandem with the Beas complex with which it forms an integrated system, it irrigates over 3.5 million ha of land in Punjab, Haryana and Rajasthan, and generates some 13,000 million units of energy per annum. The resultant agricultural and industrial production has an annual payback considerably greater than the Rs 1050 crores initially invested in this hugely successful, interlocking multipurpose project.

It would be inconceivable to think of an India minus the Bhakra-Pong complex today. Its waters provide seven to eight food-grains out of ten in every mouthful consumed by all those dependent on the public distribution system in any part of the country. The grain procured from the area irrigated by it in Punjab and Haryana alone would be more than double the 100 million tonnes or more of PL 480 supplies for which India was dependent on the United States for all of the third quarter of this century. Bhakra-Pong has given meaning to independence and virtually relegated the spectre of famine to history. It would be extravagant to suggest that the battle against hunger has been won. But the contribution of Bhakra-Pong to what has admittedly been achieved is striking and self-evident.

Beginnings

Irrigation came ,early to the Indus Valley and sustained a proud civilisation. Simple inundation canals gave way to a large new modern irrigation system in the latter 19th and first half of the 20th century. The Upper Bari Doab Canal, between the Ravi and the Beas (1859), and the Sirhind Canal in Southern Punjab (1872) were among works benefitting what is East Punjab today. Diversions, links and barrages continued to be developed to irrigate 1.02 m ha. of land in Punjab, Sind and some of the adjacent princely states by 1947 with 73 MAF being diverted from the Indus. There was no storage on this entire system until the Bhakra Dam on the Sutlej began filling to a limited extent in 1958. (Gulati, N.D., 1973)

The so-called canal waters dispute between India and Pakistan that followed the partitioning of this unique, integrated irrigation system came to a head in 1948. With the good offices of the World Bank, an overall settlement was arived at with the signing of the Indus Waters Treaty in 1960. Broad agreement on the division of the waters of the Indus system had been reached much earlier on the basis of historic uses, devlopment plans and future needs. The Treaty apportioned the entire waters of the three eastern rivers, the Sutlej, Beas and Ravi to India and those of the three Western rivers, namely, the Indus, Jhelum and Chenab (less some marginal uses in Jammu and Kashmir) to Pakistan. The division was roughly in the ratio of 19 : 81 in favour of Pakistan. The Indian allocation of 34 MAF was, however, now exclusively available for it to utilise in whatever manner it deemed best. The story of the negotiations has been well told by Gulati The Bhakra-Beas complex was the kingpin of the Indian plan for harnessing these waters.

The origin of the Bhakra project dates back to the early years of this century when southeast Punjab was, unbelievably to present generations, prone to periodic drought and famine. The idea of a storage dam on the Sutlej was mooted by a former Punjab Governor, Sir Louis Dane, who was struck by the possibilities of closing the gorge he had noticed while touring the area in 1908. A detailed report and design estimate was prepared in 1910. A cost of Rs 3.72 crores was considered prohibitive and the proposal was shelved as unremunerative. The idea was revived a few years later and a 395-foot high dam was proposed in 1919 to impound 2.58 MAF of water. There was no provision for generating electricity. Priority was , however, accorded to other Sutlej valley diversion projects and it was not until 1927-32 that a 500-foot high dam was again proposed.

In 1939 work on a detailed project report was initiated to store 4.75 MAF of water and install a hydel capacity of 200 MW. This would have entailed a maximum reservoir level of 1600 feet above sea level which would have submerged Bilaspur town in violation of an agreement between the Pubjab Government and the Raja of Bilaspur. Hence the design was altered to keep the maximum reservoir level at 1580 feet. With partition, the situation changed and it was decided to build a dam to the safe maximum optimal height which gave a full reservoir level of 1680 feet, later raised to 1685 feet. (BBMB, 1988) The revised project for a 680-foot high straight gravity concrete dam with a gross storage of 7.80 MAF was ready by 1951, preliminary works having commenced in 1948.

Bhakra-Beas Components

Work on the Bhakra Project continued *pari passu* with negotiations on the Indus Waters Treaty and the conceptualisation and subsequent designing of the component units and links to produce an overall plan for the harnessing and utilisation of the eastern rivers. As the main dam would obviously take time to complete, work was hastened on the Nangal dam and the Bhakra canal system offtaking from it so that early irrigation benefits and some power from the Nangal canal falls should be realised. These were completed in good time and the Bhakra canals started running in stages from 1954. The Pong dam was approved in 1958 and the other major component, the Beas Project, began to take shape. The Thein Dam on the Ravi was also contemplated, but in the absence of funds was only taken up later and is still under construction.

The Bhakra Dam, completed in 1963, and the subsidiary Nangal barrage 12 kilometres downstream, work in tandem. The Nangal barrage operates a balancing-cum-storage reservoir for the waters released through the Bhakra turbines and provides the headworks for the 14,500 cusec Nangal hydel channel with hydro stations at two falls at Ganguwal and Kotla with a combined installed capacity of 154 MW. The Bhakra Main Canal (12,500 cusecs) takes off from the Ropar headworks 60 km from Nangal Dam, to irrigate large tracts and firm up supplies to some older systems.

The Bhakra left bank power house, completed in 1961, houses five units of 90 MW to give an installed capacity of 450 MW, subsequently uprated to 540 MW as a result of technological improvements. It had initially been thought that load development would be gradual. Events belied this forecast and with demand picking up rapidly, the right bank power house was expedited and completed by 1969. This has 5 × 120 MW units, subsequently uprated to 132 MW each, to give an installed capacity of 660 MW.

Work had, meanwhile, commenced on the Beas Project which comprises two units. Unit-I, the Beas-Sutlej Link, is essentially a power project and diverts 3.82 MAF of Beas waters at Pandoh, 113 km upstream of the Pong Dam, into the Sutlej over a 150-foot drop. The Dehar power house at this point has an installed capacity of 990 MW, the tail race waters then flowing down the Sutlej to be stored in Bhakra's Gobindsagar reservoir. The diversion from Pandoh to Dehar is through an open channel and three tunnels with a combined length of over 25 km, an intricate job of engineering in itself.

Unit-II is the Pong Dam on the Beas, just before it enters the plains at Talwara, with a gross storage of 6.95 MAF behind a 435 feet earth-core gravel shell dam. The power plant at the base of the dam has an installed capacity of 360 MW. Unit-II is a joint venture undertaken by Punjab, Haryana and Rajasthan.

The Pong site was originally in what was Punjab's Kangra district. But after the reorganisation of Punjab and Himachal, it straddles the new inter-State border with the entire reservoir and the Pandoh and Dehar sites lying in Himachal. Likewise, the Gobindsagar reservoir backs into Himachal. This configuration initially excluded Himachal from participation in the Bhakra-Beas Management Board, headquartered at Chandigarh, which operates the project for the benefit of the partner States of Punjab, Haryana and Rajasthan. It was later proposed that Himachal be treated as part of the BBMB on par with the other three States with an appropriate reapportionment of its capital and operational costs. This organisational hiccup needed resolution as catchment area treatment, tourism and fisheries in the two reservoirs vests in Himachal which also now carries the political burden of ensuring the final resettlement in the IG Nahar command in Rajasthan of the Pong Dam oustees displaced from Kangra.

When one thinks of current-day costs, the outlay on the Bhakra-Beas project must seem absurdly small. The entire Bhakra-Nangal project with canal systems and power houses cost just Rs 245.28 crores, admittedly a tidy sum at the time. While the Bhakra left and right bank power houses cost Rs 98.30 crores, the uprating of the turbines to add 150 MW capacity was accomplished for under Rs 10 crores as against an investment of approximately Rs 2.5 to 3 crores per MW of installed capacity that would be required today. Further uprating of the right bank power house from 132 MW to 157 MW per turbine or an addition of 125 MW in all is programmed to be undertaken within the next few years. The Beas-Sutlej Link cost Rs 449.17 crores and the Pong Dam Rs 325.88 crores. This totals no more than about Rs 1032 crores. The transmission system constitutes a separate investment.

The other main component of the Eastern (Indus) Rivers complex is the Thein Dam on the Ravi above Pathankot. This 160 metre high structure will store 2.19 MAF of water in the Ranjitsagar Lake spreading over 90 square km. It will have 4 × 250 MW units and is expected to generate 1509 million units of energy. The project was designed in the early 1960s and was estimated to cost around Rs 300 crores. It was, however, only cleared in 1982 but suffered a major setback in the 1988 floods when the diversion tunnels collapsed. The Rs 1566 crore project (1991) is now expected to be completed in 1997 and will irrigate 348,000 ha, apart from storing waters for the IG Nahar project along with the Pong Dam. Releases through the turbines will be trapped 11 km downstream behind the 60 metre high Shahpurkandi dam with a hydel capacity of 168 MW. This dam, recently approved, will cost Rs 756 crores at 1991 prices and, being situated on the Punjab-J&K border, will irrigate some high lands in Jammu which cannot be commanded by the Madhopur barrage, headworks for the Ravi-

Beas Link, 8 km below. Finances for this project are yet to be arranged. One of the problems about expediting work on the Thein Dam has been anxieties on part of the personnel engaged about future employment once the project is completed. Sanction for the Shahpurkandi dam might therefore have a salutary effect on the work-ethic at Thein Dam.

Rehabilitation Measures

Any activity requiring land acquisition in settled areas entails displacement of populations. This is a genuine hardship that needs to be more than recompensed in material and emotional or psychological terms. Bhakra and Pong represent early learning experiences.

The Bhakra reservoir submerged 44,153 acres of land, affecting 371 villages with a population of about 36,000 persons or 7206 families. Bilaspur town, with a population of around 4000, was also submerged. Of the 7206 affected families, 5027 were given cash compensation under the terms of the Land Acquisition Act and resettled elsewhere on their own. A Bhakra Rehabilitation Committee was appointed to examine the future of the remaining 2179 families. Since these farmers expressed a desire for resettlement in the Bhakra command, the Committee located suitable lands in compact blocks in Hissar and Sirsa districts in Punjab and Haryana respectively and 13,200 acres were accordingly acquired in 30 villages. As the amount due in compensation exceeded the value of the lands acquired, the displaced families were compensated partly in land and partly in cash, subject to a ceiling of 25 acres but with the guarantee that none would be allotted less than the area of his farmlands acquired if the amount of compensation was adequate to meet the cost. Further, it was stated that no family would receive less than five acres in compensation.

Landless tenants were declared eligible to be given an area equivalent to their submerged tenancy as mentioned in the revenue records for the 1957 kharif season, subject to a maximum of five acres, the price to be recovered in 20 half-yearly instalments at 5.25 per cent interest and with a two year grace period. Other labourers and artisans from the affected villages were offered a half-acre homestead plot free of cost in the new abadi villages in

Hissar and Sirsa if they desired to shift. The usual amenities were provided at the new abadi sites and those moving were assisted with housing loans, rail or bus fares, a rehabilitation allowance and assured canal water. Those choosing to remain in Himachal were granted free fishing licences in Gobindsagar for three years and assured new ferries and roads in place of those submerged or dislocated, gainful employment in the Bhakra project with training for skilled jobs, and additional amenities in their new places of residence. By the end of 1991, proprietary rights had been vested in most of those resettling in Hissar and Sirsa. A few cases were delayed on account of allottees not visiting the office of the concerned official at Hissar to execute the conveyance deeds and complete other formalities. (Anon, December 1991)

The story of the Pong dam oustees is narrated in the chapter on the IG Nahar and need not be repeated here. Those displaced by Unit-I, the Beas-Sutlej Link, and some of those affected by Unit-II, the Pong Dam, also utilised their compensation to resettle locally in higher reaches. In the case of Pong Dam, while the designed maximum reservoir level is 1430 feet on the basis of what is termed a 10,000 year flood, land was acquired only up to an elevation of 1410 feet, in view of the low risk involved for those living above that level, in order to avoid exececessive displacement. Indeed, since storage commenced in 1974, the reservoir has only filled up to the designed 6.95 MAF on three occasions.

The submergence area under Ranjitsagar which will form behind the Thein Dam is much smaller. The original rehabilitation programme of Rs 4 crores is now estimated to cost Rs 86 crores on account of escalations caused by long delay. Here again, Punjab has to pay compensation to persons displaced in Himachal.

Problems arise whenever the affected area and the benefited area fall in two or more States. In the case of the Bhakra-Beas Projects, Himachal was only marginally affected initially and the main burden of submergence and displacement fell on Punjab, except in the case of the Beas-Bhakra Link, as Pandoh and Dehar were always in Himachal. However, with the reorganisation of Punjab in 1967, Haryana came into being as a new beneficiary State. With the transfer of Kangra District to Himachal, Punjab became almost exclusively a beneficiary State while Himachal emerged as virtually the sole affected State without adequate compensation for having to carry this unexpected burden. Rajasthan, of course, continued as a beneficiary. Thus the Pong Dam oustees belong to Himachal but the resettlement areas lie in Punjab, Haryana and Rajasthan. In the case of displaced persons too Himachal feels aggrieved that electricity to some Himachal villages in which Bhakra oustees settled on their own did not receive Bhakra power for up to 18 years. More imaginative planning and coordination between the concerned States and a revised cost-benefit sharing formula that took account of Himachal's new status might have averted much wrangling and heartburn all round and ensured smoother progress. Responsibility, however, does not lie on the Project as much as on the structures of inter-state management and discourse. The failure of the Government of India to set up an inter-State Council until recently or to use the mechanism of the now-defunct Zonal Councils contributed to this unhappy situation.

Sedimentation and Catchment Area Treatment

The same problem of divided responsibilities has affected other issues. Soil conservation and siltation is one such.

The Ravi, Beas and Sutlej have substantial upper catchments, the Pong catchment extending over 12,560 square km. The Sutlej catchment of 56,980 sq. km. is much larger, this being an "antecedent", trans-Himalayan river rising near Lake Mansarowar in Tibet. The Himalaya is a fragile mountain, still pushing upwards under tectonic impulses. Seismic shaking and weathering cause a great deal of erosion with the debris that is swept into rivers and torrents being transported to the plains and ultimately to the sea. Of the approximately 27,000 acre-feet of silt flushed down the Sutlej annually, as measured near Bhakra, some 60 per cent is generated in Tibet and the balance within Himachal. Dams trap sediment, especially the heavier particles, which gradully begin to eat into its storage capacity. All dams therefore provide for what is termed dead storage, leaving a dependable live storage avilable for benefical use over the designed life of the dam.

The Bhakra Dam was originally estimated to have a life of 630 years based on an assumed annual rate of siltation of 4.29 ha-metre per 100 square km. The siltation rate was, however, observed to be 8.14 in 1964 but had declined to 6.22 by 1979.

(National Land Use and Conservation Board, 1988) The life of the dam has been reassessed at around 380–400 years but is expected to improve with appropriate soil conservation measures and catchment area treatment.

A higher rate of siltation than assumed is not uncommon in Indian water resource projects and may be attributed to many factors. Critics argue that it is typically because engineers and planners underestimate problems and costs and overestimate benefits in order to ensure approval for their pet projects. While there could be an element of truth in this, the point should not be exaggerated. Sediment recordings have often been limited at the stage of project formulation as many catchments are in remote areas. With the construction of the project the upper catchments become increasingly accessible. This makes for improved measurements. The construction phase and the activity and movement this generates, including the making of roads and tunnels, blasting, and the very opening up of the upper catchment stimulates economic activity. Other developments such as the construction of more (border) roads, logging, quarrying and so on do disturb the earlier serenity and destabilise slopes and can trigger landslides. It may take quite some time before balance and stability are restored and it is precisely during this period that the new sediment readings are registered.

Averages will also vary with the periods recorded and natural events such as large earthquakes, floods and bursting of debris dams created by major slides which might be included or excluded. The impact of increasing population and livestock on environmental degradation through cultivation of marginal lands on steeper slopes, over-grazing, and lopping and topping for fuel and forage must also contribute to greater erosion. However, with appropriate catchment area treatment conditions can and often do improve and it is for this reason that such programmes are now mandatory in all river valley projects.

The Gobindsagar lake started filling 1959 and the first sedimentation survey was conducted in 1963. The percentage loss of dead and live storage capacity up to the end of 1988 worked out to 21.99 and 6.42 per cent respectively with an overall capacity loss of 10.26 per cent. (BBMB, March 1990) The total silt deposition in Gobindsagar in 28 years up to 1987 was 0.935 MAF, the average up to 1986 working out to 25,300 acre-feet per annum as against the design figure of 27,250 acre-feet. The trend line showed a curve that rose above the assumed rates of siltation in the initial years and then flattened and fell below the estimated design figure. Further improvement may be expected with soil conservation measures. But what is worrying is the configuration of the silt that has already been deposited in the reservoir. The normal expectation is that as a flowing river enters the reservoir it will shed its sediment load at the deep end within the dead storage forming a delta as when a river meets the sea. However, the actual geometry of silt deposition has tended to describe a hump rather than a delta, thus constituting a subsidiary dam inhibiting passage of as much silt as anticipated through the dam. (BBMB, October 1988) In over 30 years the delta has been moving at the rate of half a kilometre per annum towards the dam.

The sixth sedimentation report for the Pong Dam, 1988-89, indicated an average sedimentation rate of 26,500 acre-feet per annum as against the designed figure of 20,500 acre-feet. The total silt deposited in the reservoir between 1974 and 1989 worked out to 0.3975 MAF or 5.72 per cent of the gross storage capacity. The small Pandoh reservoir, however, had accumulated 16,886 acre-feet of silt as against a gross storage capacity of 33,240 acre-feet as of October 1989. (BBMB, March 1990)

In the case of both Bhakra and Pong, a catchment area treatment plan was evolved in two parts. The BBMB is responsible for treatment and plantations on a certain "fringe" area along the rim of the two reservoirs. This was to be done on the basis of a "tree patta" programme to be prepared by the Himachal Forest Department to ensure popular paticipation by and cooperation from the local people in protecting the plantations, as they would enjoy usufruct rights. While this programme is to be funded by the BBMB, treatment of the larger catchment was to be taken up by the Himachal authorities with Central assistance.

At Bhakra, the fringe area constitutes a band between 1680 and 1700 feet elevations around Gobindsagar. In respect of the Pong reservoir, the designated fringe area lies between 1410 and 1431 feet elevations. In 1981 a committee on afforestation and soil conservation measures in the Bhakra and Beas catchment areas recommended a five year work plan costing Rs 14 crores, including an outlay of Rs 2 crores for the fringe areas. The Departments of Forests, Roads (Public Works), Agriculture and Animal Husbandry of the Himachal government were involved. Owing to lack of funds it was subsequently decided that a good part of the work might be undertaken through the Centrally-sponsored programme for Integrated Watershed Management in the catchments of River Valley Projects. The assigned tasks included survey of 250,000 ha of priority micro-watersheds, afforestation of 10,000 ha, treatment of 10,000 ha of grazing and pasture land, treatment of 1500 ha of agricultural land, and execution of minor engineering works like gully plugs and check dams over 3300 ha, all of these in "high to very highly eroding areas" or "critical areas." However, no funds could be made available during the Sixth or Seventh Plans while the cost estimates for the original Rs 14 crore programme was revised upwards to Rs 42 crores in 1987. The BBMB pleaded that critical areas should be treated on a "war footing" to prevent excessive siltation in the two reservoirs. It estimated that up to 45 per cent of the sub-watersheds in the Sutlej and Beas catchments would fall in the "very high" and "high" erodibility class for which the Centre had been able to allot no more than Rs 2 crores with which only 4000 ha could be effectively treated. The Rs 42 crore programme included treatment of 200 km of road, compensation for stoppage of stone quarrying, fencing of portions of the reservoir peripheries, afforestation of the periphery areas, and an intensive cattle development programme.

Everything said, little has been done. The Himachal administration pleads lack of funds while others complain that the State has little interest in greater expenditures or effort as it is not a major beneficiary and has low stakes in the Bhakra and Beas projects. Nor are the local people cooperative. They have been resentful of plantations, uproot the trees in the fringe areas as they prefer to cultivate this strip and indeed farm the drawdown areas as the reservoir recedes after mid-December. The BBMB for its part does not favour drawdown cultivation as this destabilises the reservoir rim and, by loosening the soil, triggers landslides and aggravates siltation in the reservoirs. The BBMB has banned drawdown cultivation but farmers continue to do so unauthorisedly and are willing to pay damages equal to double the land revenue, as decided by the Bhakra Rehabilitation Committee. The penalty is, however, notional as the civil authorities are unwilling to make recoveries. Meanwhile, the cost of catchment area treatment is rapidly mounting and unofficial estimates now reportedly place this in the region of Rs 800 crores.

Dredging silt from the depths of a large and deep reservoir is not easy and would be very expensive and pose a problem of disposal of the spoils. However, the BBMB has resorted to dredging on a small scale in the balancing reservoir of the Dehar power station which is part of the Beas-Sutlej Link. The balancing reservoir has a capacity of 3000 acre-feet. Some 500-600 acre-feet is sought to be dredged annually and the silt disposed of in nearby nullahs. But this ultimately flows back into the Pong reservoir, though the Dehar power house is undoubtedly benefited.

There is every reason to expect that if serious efforts are made to pursue catchment area treatment and afforestation plans there will be a positive impact on dam siltation. The construction of upstream storages, such as the proposed Kol dam on the Sutlej above Dehar with an installed capacity of 600 MW, and lesser check dams would provide silt traps that would reduce the sediment load entering the Bhakra and Pong reservoirs. These programmes must be given far higher priority in order to protect the life and efficiency of existing assets of great national value. Problems of inter-state coordination are easily soluble and only require a bit of political will. Worries about the financial burden would also ease if these programmes are dovetailed into wider watershed management programmes that have soil conservation and afforestation as only one element within a larger effort to conserve and improve the national resource base and ameliorate the lives of the local hill communities through income generation and employment. Watershed management programmes are an important pre-investment in water resource development.

Seen this way, one should be looking at Himachal as a whole and not at just the Bhakra and Beas catchments. Indeed, Himachal shares the Ravi, Chenab and Yamuna catchments too.

There is a component of natural mass wastage which is beyond human control. The smaller, though significantly large, element of manmade interventions is, however, something that can and must be addressed. As in many other parts of the country, population and livestock growth have put tremendous pressure on the land in excess of its carrying capacity for sustainable use. Himachal's population was approximately 3.15 m after the current reorganisation in 1967. It had risen to 5.1 m by 1991. In 1967 there were some 4.20 m head of cattle and 0.81 m goats in the State of which nearly 2.37 m cattle and an unspecified number of goats grazed and browsed in the forests, impeding regeneration. By 1977 the number of goats had gone up to 1.03 m. The lopping of trees like oak for fodder and bedding is destructive and, ultimately, killing. Firing of forests by herdsmen to secure a fresh flush of grass has caused great damage. During 1955 and 1966 as many as 2340 such fires were reported resulting in damage to 98,000 ha of forest according to official statistics. The movement of cattle up and down hill slopes also loosens the soil. Road-building (with blasting), quarrying and tunnelling for water resource projects have also variously encouraged erosion.

Population growth has also resulted in diversion of land to agriculture, much of which is rainfed. Crop farming on unterraced slopes or poorly maintained marginal terraces is ecologically unviable. Himachal has fortunately switched to a more scientific pattern of hill farming and horticulture with the area under fruit doubling from 109,000 acres in 1970-71 to 228,000 acres in 1981-82. Till recently forests were being heavily cut to manufacture apple boxes. Fortunately the technology of corrugated cardboard crates has developed and the wasteful use of crate-wood is being gradually overcome.

The purpose of this digression is to underline the importance of land use planning generally and to avoid the simplistic belief that water resource projects are the major cause of deforestation and erosion and are thereby both indirectly and directly responsible for floods.

The Floodgate Fallacy

The popular mythology about the causal link between deforestation and floods has been discussed elsewhere in this volume. Another widely held belief is that spillage through dams when a reservoir is full causes floods. Poor reservoir management can be damaging but references to "opening the floodgates" must normally be taken as a literary allusion rather than as a literal fact.

The charge that the BBMB caused the disastrous floods in Punjab in 1988 by opening the floodgates of Bhakra and Pong was made by some critics, including Khalistani militants who, on that pretext, tragically gunned down the then BBMB Chairman, Major-General B.N. Kumar. The militants obviously had other motives and the Bhakra system had earlier been targeted by them with sabotage attacks on the Main Canal. The charge levelled was that having filled the Gobindsagar and Pong reservoirs to capacity, the BBMB "criminally," permitted huge discharges from the dams without warning. (Shiva, Vandana. 1991)

What are the facts? The Sutlej Valley could previously absorb a flood of 450,000 to 500,000 cusecs. Following the construction of the Bhakra Dam, the river was jacketed downstream within embankments, reducing its flood capacity to about 200,000 cusecs. The bed within a river's natural flood banks is not normally acquired and, with the protection afforded by the dam, is in fact cultivated by Punjab farmers.

The Bhakra dam is essentially an irrigation project committed to making timely water supplies to 2.6 m ha of farmland. Its operations are therefore designed to avoid spilling water, and all releases are passed through its turbines. The BBMB makes discharge forecasts based on snow-melt and related studies through remote sensing data and regular meteorological transmissions. These are then compiled in a flood forecast to regulate reservoir operation during the filling season from May 21 to September 20 for Bhakra and June 21 to September 20 for Pong.

The BBMB is guided by two parameters: to achieve maximum filling by September 20, and ensure the safety of the dam in accordance with clearly specified conditions. The filling and discharging of the reservoirs is jointly reviewed at a monthly Technical Committee meeting at which chief engineers of the partner States' irrigation departments and State electricity boards are present. The meteorological forecast for September 1988 indicated a 19 per cent deficiency in rainfall and the expectation therefore of about 100 mm of rain was confirmed by subsequent inflows and weather conditions. The normal filling levels of the Bhakra and Pong reservoirs are 1685 feet and 1400 feet respectively and these levels had been reached by September 13. The heavens opened thereafter and, instead of an anticipated 110 mm of rainfall in the Sutlej catchment during the month of September, as much as 436 mm was actually received, almost all of this between September 23 and 26. About 50 per cent of the normal annual precipitation was experienced during those four days, with a cloudburst of 85 mm of rain in 45 minutes on September 26 in a 20-30 km band along the Thein-Pong-Bhakra axis. Bhakra moderated a 318,000 cusec flood by absorbing 176,000 cusecs. The rest was released but coincided with a far larger discharge of 478,000 cusecs in the rivulets and streams, especially Swan nadi, astride the axis of the cloudburst, which was disgorged into the Sutlej below Bhakra, a little above Ropar. Likewise, Pong Dam moderated a flood of 301,228 cusecs on September 26 to 261,000 cusecs while the Sohan nadi , Chakki river and other rivulets discharged almost as much into the Beas below the dam. The Pong reservoir was actually filled up to 1494.70 feet elevation to absorb 230,000 cusec-days during the most critical period of September 25-26. Further filling of the two reservoirs could not have been permitted without endangering their safety. As a mater of fact, the Bhakra Dam tilted by 0.95 to 0.99 inches as against a designed "non-seismic tilt" of 1.03 inches. Landslides aggravated the problem and the Pandoh reservoir suffered heavy silation. (BBMB, 1988)

While it is easy to accuse the BBMB of opening the floodgates, the evidence clearly shows that were it not for the cushion provided by the two dams, the floods would have been even worse. The same story repeated itself in the great floods of 1993 though the same kind of blame was not sought to be attached to the Bhakra-Beas projects. Patiala town was flooded by Patiala ki Roa nadi which drained 465 mm of rainfall in just 36 hours from July 10. There were unprecedented floods over large areas of the United States mid-west in the summer of 1993, with the Mississippi and Missouri overflowing their banks. Extraordinary and persistent rains brought on the deluge. Dam-builders have never aspired to playing God.

Bread Basket

At the time of Independence, (East) Punjab was the agriculturally poorer part of the undivided State and what is now Haryana, the poorer half of that residual Punjab. Bhakra effected the difference seen today, bringing 2.6 m ha under new or improved irrigation, recharging groundwater in what was a largely semi-arid region, and providing the electricity to lift water economically through a rural electrification network.

The figures tell their own story. In 1960-61 total foodgrains production in what is today Pubjab was 3.16 m tonnes. By 1990-91 it had gone up to 19.25 m tonnes, with a tripling of both rice and wheat yields. There was, however, a diversion of acreage from coarse grains, pulses, gram and oilseeds to paddy and wheat which have since remained the dominant and most profitable rotation cycle, especially after the green revolution with its emphasis on improved seeds and larger use of fertilisers and pesticides with increased irrigation. The combined use of N, P and K fertilisers rose during this same period from 5000 tonnes to 1.22 m tonnes. The production of cash crops like cotton and sugarcane rose appreciably. Almost 94 per cent of Punjab's sown area is under irrigation which now has an intensity of about 176. The record in Haryana is also impressive. Foodgrain production rose from 2.6 m tonnes in 1966-67 to 9.6 m tonnes in 1991, and there were very marked increases in oilseed and cotton production, in the consumption of fertilisers and in the area brought under irrigation. An interesting aspect of farming in Haryana is irrigated tree cultivation. The State was deficit in timber until 1987 but now exports over Rs 100 crores worth of timber and wood products (like fruit boxes) to other parts of the country.

Another index of agricultural advancement is mechanisation. Punjab had 773,000 tubewells in 1991, nearly four-fifths driven by electric pumps and the rest by diesel engines. 265,000 tractors and 195,000 tillers were reported in use and 5000 combine harvesters, 3100 reapers, 297,000 threshers, 100,000 seed-cum-fertiliser drills and 435,000 spray pumps. Haryana in 1991 had 107,000 tractors, and 458,000 tubewells in operation. The State supplied 2.76 m tonnes to the Central food reserves in 1991 while Punjab's contribution to the public distribution system in percentage terms was 61 for wheat and 43 for rice. This is truly India's bread basket.

Both Punjab and Haryana make exaggerated claims to more water than is ever going to be available. While both are more economical users than farmers in many other States, water use efficiency is capable of improvement. Consolidation and warabandi have helped. Both states are modernising their irrigation systems with lining of canals and water courses, installation of control structures for better regulation, and introduction of sprinkler systems. Water user associations need to be further developed and canal water rates are clearly low. Roughly 58 per cent of irrigation comes from tubewells, including state-owned deep tubewells and augmentation wells. The power subsidy here is being sought to be gradually reduced by raising the electricity tariff. The formal view is that the farmer cannot afford to bear higher rates as procurement prices are too low. But an economic price for water and power will stabilise, not harm, agriculture and scientific irrigation in the long run.

Haryana's southwest plateau borders the desert. The State has laid great stress on lifting water 200/400 feet to irrigate these tracts through five lift schemes which will require more water and power for their projected extension. This is why Haryana sets store on the Sutlej-Yamuna Link canal which has become such a bone of contention with Punjab in terms of allocating the surplus waters of the Ravi and Beas. It has, however, sought to regulate its flood waters by developing intermediate storages and plans to harvest the sub-montane nullahs as a soil-cum-water conservation measure as at Sukhomajri, near Chandigarh.

Punjab has similar conservation programmes and plans to link a number of low dams over foothill torrents or choes through the Kandi canal, offtaking from the Shah Nahar barrage near Talwara. It will reach Hoshiarpur by 1995 and will be subsequently extended to Balachur. This will benefit some 55,000 ha of undulating, gullied kandi lands and moderate floods and erosion.

The SYL Canal is an emotional issue linked to the inter-State dispute between Haryana and Punjab about water allocations. The 175 km Haryana portion of this 196 km, 7500 cusec canal, originates from the tail of the Anandpur hydel channel and will provide Punjab with 200,000 ha of irrigation and 55 MW of power on a canal fall. The revised cost of the project is around Rs 570 crores. But the capacity of the Bhakra Main Branch in Haryana has declined by 1000 cusecs and that of the Narwana branch, which links up with the Western Jamuna Canal, by 622 cusecs, both of these for lack of repairs. If this were expeditiously repaired, now that the BBMB has completed the difficult task of restoring the Bhakra Main which was damaged by terrorists, Haryana would be able to draw a substantial quantum of water relatively quickly and cheaply. The Bhakra Main repair would cost Rs 8 crores and carry an additional 700,000 cusecs of water to Haryana. Politics has intruded all round and both states are committed to more extravagant options.

Green Revolution Indicted

Despite the problems recounted, many of them amenable to correction, there is little doubt that the green revolution has transformed Punjab and Harayna and that this would not have been possible but for Bhakra-Pong. Deprived of their waters Punjab and Haryana would have remained semi-arid tracts except for some modest and uncertain inundation irrigation. Without the recharge from widespread surface irrigation, the extensive groundwater development now evident would have been impossible, with the water table being deep, quantitatively limited and brackish over large parts of southern and soutwestern Punjab and Haryana. Nor would cheap power have been available to extract such groundwater as was available with the facility with which this is now done. Punjab and Haryana would not have been the nation's granary nor enjoyed the highest per capita income in the country.

There is, however, an alternative view now being projected. The "gift" of Bhakra-Nangal has been the Punjab problem itself because of its alleged outcome, namely "soil death and terrorism" (Sahgal, July 1993). "So strong is the brain-washing (about the virtues of large dams) that even well-meaning people often argue that without the Bhakra Dam, Punjab would not have had its green revolution, without which India's growing population would have faced starvation. The last statement is untrue. Today the green revolution is in truth proving to be the greatest threat to the food security of the nation. . . . The chemical-intensive, energy-intensive and water-intensive monocultures of hybrid crops have been literally leaving behind a trail of poison, plague and plunder". (Mansata, June 21, 1992)

Another critique is to be found in the film *The Good Earth* produced by Krishnendu Bose and sponsored by INTACH. "All

I have tried to prove is that the green revolution is turning out to be a farce", says the filmmaker to *The Pioneer* whose reporter echoes this sentiment: "Is it not true that the green revolution, which has now begun to show its catastrophic consequences, was a predictable sequence in Punjab's big boom, and a possible origin for the birth of right-wing terrorism, secessionism?" (Sengupta, July 8, 1993) What Bose says he shows in his film is that organic farming offers an alternative "eco-friendly conception of life which has existed for centuries in our village and tribal societies".

A far more elaborate indictment of the green revolution and Bhakra comes from Vandana Shiva (1991) who argues that after two decades of the green revolution "Punjab is neither a land of prosperity nor peace. It is a region riddled with discontent and violence. Instead of abundance, Punjab has been left with diseased soils, pest-infested crops, waterlogged deserts, and indebted and discontented farmers. Instead of peace, Punjab has inherited conflict and violence". The green revolution is seen by her as a techno-political strategy promoted by the West to counter the threat of communist insurgency and peasant revolutionaries: "The ecological and ethnic crisis in Punjab can be viewed as arising from a basic and unresolved conflict between the demands of diversity, decentralisation and democracy on the one hand and the demands of uniformity, centralisation and militarisation on the other. Control over nature and control over people were essential elements of the centralised and centralising strategy of the Green Revolution. Ecological breakdown of nature and the political breakdown of society were essential implications of a policy based on tearing apart both nature and society."

The miracle seeds, she continues, demand far more water than traditional varieties, and increasing doses of fertiliser and pesticides to combat new plant diseases and infestations. This large requirement of water has led to widespread waterlogging and salinisation on the one hand and manmade disasters when the Bhakra and Pong floodgates have been opened on the other. Canals constructed across the grain of the land, including the ongoing SYL Canal, have and will create drainage problems. The emphasis on high-yielding varieties has resulted in a shift of acreage from oilseeds, pulses and coarse grains to rice and wheat which are water-guzzlers. Protective irrigation, which was more equitable, has yielded to intensive irrigation which is less productive in terms of output per unit of water consumed in relation to traditional rice and wheat varieties. Crop diversity has yielded to homogenisation of a few gene-types, monocultures and loss of bio-diversity.

The new paddy-wheat cropping cycle is soil-depleting and chemicalisation has led to reduced use of organic manures and micro-nutrient deficiencies. Soil toxicity has created diseased and dying soils.

Reliance on expensive surface drains and conjuctive use of water to provide vertical drainage have not really helped, she contends. The land slopes towards southwest Punjab which is waterlogged and has brackish aquifers. Hence only canal water can be used. The skimming-well technology recommended here, which entails siphoning off a top layer of fresh water without pumping the underlying brackish water strata, is expensive. Elsewhere too, excessive groundwater extraction has caused a sharp fall in the water table. Says Shiva, "it is estimated that an irrigation pump powered by a 7.5 kg (*sic*) electric motor takes five hours and one man to irrigate an acre of wheat compared to 60 bullock-hours and 60 man-hours with a Persian wheel. The calculation that was never carried out by Green Revolution experts was that the Persian wheel has supported agriculture for centuries while the energised pump threatens to dessicate large areas of prime farmland in less than two decades". Likewise, other forms of farm mechanisation, such as tractorisation, have displaced labour and depressed real wages. Debt has increased and small farms have been rendered unviable.

This elite, urban exploitation of Bharat is a product of centralisation with the BBMB, a centrally-managed structure, at the polar end of de-peasantisation of the peasantry. It has engendered water conflicts between classes of farmers, between Punjab and Haryana and between the Centre and these States. Thus large dams and high water demands have led to "major ecological impacts on the water balance and political impacts on the power balance of the region". The promise of a "second green revolution" through bio-technology, new miracle seeds for tomatoes, potatoes and the rest, the patenting of seeds through Dunkel protocols and Pepsi, threatens the Third World with "recolonisation". The false and short-lived prosperity generated by the green revolution has caused cultural and social destabilisation. "Circulation of new cash in a society whose old forms of life had been dislocated led to an epidemic of social diseases like alcoholism, smoking, drug-addiction, the spread of pornographic films and literature, and violence against women." An alienated and disoriented Sikh peasantry and educated unemployed, finding an "ethical vacuum", were attracted by religious revivalism which crystallised in the emergence of a separatist Sikh identity. Bhindranwale attracted attention by crusading against cultural corruption and "degenerative consumption" in Punjab. It was inevitable that this should be perceived as a conflict between "a communal Hindu Centre and Sikh people in a Sikh State".

So terrorism and violence in Punjab, according to Shiva, can be directly traced to the green revolution. "Restoring a decentered (*sic*) and ecological system of food production is the only strategy for creating a lasting peace". Simple. QED.

Bittu Sahgal endorses this thesis. Bhindranwale "preached violence in retaliation for the (green revolution) plot against Punjab by the Centre". Hence, "the Punjab problem will never be solved till it is recognised that its roots lie in the restoration of environmental and political stability". This in July 1993 by when the law and order situation in Punjab had dramatically changed. Sahgal's solution: organic farming, not fresh "predatory" inroads by those threatening to "mine" Punjab's soils by growing and exporting potatoes and tomatoes "leaving behind a lethal legacy of pesticide poisoning and further soil death".

Exaggerated Fears

Organic farming is unexceptionable and has grown in vogue ever since Fukuoka wrote his *One Straw Revolution*. A number of Indian experiments have been reported from different parts of the country (Sengupta, Amit). They should continue and must receive encouragement and the country should learn from them. But nowhere in the world, including Japan, has organic farming so far progressed much beyond limited experiments. Meanwhile, bio-fertilisers, leguminous plants, azotobacters, rhizobium cultures and integrated pest management systems are being propagated in place of excessive reliance on chemical fertilisers and pesticides, while more toxic chemicals are progressively being banned.

With a rapid increase in the number of tubewells, the electrical energy costs of which are highly subsidised, and in the absence of any groundwater regulation or spacing laws, the water table has markedly declined both in Punjab and Haryana. Groundwater is being mined. In Punjab the gross recharge rate is estimated at 14 MAF per annum as against which the gross draft is 18 MAF. Going deeper is not the answer as submersible pumps must be used. Not only is this more expensive, but there is danger of saline pollution if any of the underlying brackish aquifers are pierced. The water table is falling at the rate of one foot per annum, increasing to three feet in drought years. Punjab has 118 blocks. Of these as many as 69 are designated "dark", which indicates over-exploitation of groundwater, while another 20 blocks are "grey" or marginally all right, leaving no more than 29 "white" blocks where the groundwater situation is satisfactory. The water table is declining in 84 blocks and rising in barely nine, and this too because the presence of brackish aquifers compels resort to surface irrigation, the recharge from which is building up an overlay of sweet water which can be judiciously tapped. Artificial recharge schemes are under consideration, including one to recharge the Lassara Nala bed with surplus canal water as a means of augmenting underground storage. Haryana also plans artificial recharge through unlined kharif channels, monsoon storage in abandoned channels and in flood drains after September, channel bunding, and laying of lateral recharge pipe networks from canal outlets.

Continuous irrigation over many years and the changeover to water-intensive crops had to result in a rising water table. Not only was surface drainage lacking but the natural drainage has increasingly been affected by road and rail construction, urbanisation and other development. Canal seepage has also been evident in unlined channels. The cumulative effect of this has been waterlogging and salinity in areas where capilliary action has brought up salts to form a mineralised encrustation on the land surface. Maintaining the water and salt balance are both important if soil health is to be preserved.

The natural drainage slope of the terrain has understandably led to waterlogging and salinity in the districts of Faridkot, Ferozepur and Bhatinda in southern Punjab and in the districts of Jind, Rohtak, Sirsa, Hissar and Bhiwani in central and south-western Haryana. In 1984 the area affected in Punjab was estimated to be about 200,000 ha and this had approached 300,000 ha by 1991 and is likely to increase to 600,000 ha by 2010 if left unattended. The corresponding figure of affected lands in Haryana is around 275,000 ha, with the water table within 1.5 m of the surface, the root zone of most crops.

Conjunctive use has provided much-needed vertical drainage over large areas which face the opposite problem of a falling water table. But areas with shallow brackish aquifers are more sensitive and require regulated irrigation and surface drainage. Canals, even up to watercourses, are being lined and surface drains have been constructed in many areas with outfalls in rivers, nullahs and depressions or tanks where brackish water fishery is being practised, as in Ferozepur. Sweet and brackish water is also being blended in some places. Experiments are now under way to construct sub-surface piped drainage to lead away water from the root zone of crops to sumps from where they can be pumped to suitable outfalls. Biological drainage through trees like eucalyptus has also been tried successfully. Gypsum is being applied to reclaim saline lands. Sprinkler irrigation is gaining ground as in Bhiwani and Bhatinda districts with cotton, oilseeds and citrus being grown in the latter area.

Rice was originally popular in Punjab as it is suited to water-logged areas. Continuous irrigation to leach surface salts can also work. But these merely treat the symptoms and are not sustainable in the long run.

Pakistan faces a similar problem in its Indus command. The rapid spread of waterlogging in the 1960s was successfully tackled by the Salinity Control and Reclamation Projects (SCARP) through the installation of a network of public sector deep tubewells. This was subsequently augmented by the spread of shallow tubewells put down by farmers. This follows the Indian pattern. Pakistan has, however, now discovered that salinity and waterlogging do not necessarily go together and that it is beginning to face an independent problem of "secondary salinity", particularly at the tail-end of canals where irregular supplies or shortfalls have led to heavy groundwater pumping by farmers to save their crops. In the absence of surface drainage, the water pumped here is often of poor or marginal quality. The problem is the more dangerous because it is insiduous, with sodicity being noticed after the damage has been done. (Vander Velde and Kijne, 1992) The answer to the problem is seen in the 300-kilometre-long Left Bank Outfall drain to the sea now under construction from Nawabshah, where a web of subsidiary drains converge.

Punjab, Haryana and Rajasthan face a similar danger and need to concert action as the Indus basin in India is, as in Pakistan, a single hydrological unit which cannot be satisfactorily treated piecemeal. A three-State study is in progress and, hopefully, an overall plan will emerge. Leading such drains into the Sutlej is not possible beyond the limited dilution capability of that much depleted river as it leaves India. Article IV (10) of the Indus Waters Treaty in effect enjoins this country not to permit pollution of streamflows or drainage flows into Pakistan beyond acceptable limits. The other option is to build a drain that empties into the desert or in depressions where it can do no harm or can be reclaimed or beneficially used to breed brackish fish or to extract salt.

There are technological answers up to a point to some of these problems of waterlogging and salinisation though they can at times be expensive even if cost-effective. The better answer lies elsewhere.

The paddy-wheat cycle, which is very widespread in Punjab and Haryana, makes a large draft on water, is exhausting the soil and provides a congenial environment for pests that damage these crops. The result is incremental demands for more water, chemical fertilisers and pesticides. It is not that yields are not rising or have reached saturation point. Plant breeders have not recently come up with the kind of "miracle" seeds they did some time ago. But while in Haryana wheat and rice yields are 3.6 and 2.8 tonnes per ha, and a little more in Punjab, the biological maximum or genetic potential is as much as eight and ten tonnes per ha.

The paddy-wheat cycle is favoured partly because the two states contribute about eight million tonnes of wheat and four million tonnes of rice to the national food reserve, which constitutes the mainstay of the public distribution system, and partly because farmers find this the most profitable rotation. This has something to do with relative farm prices which are influenced by official procurement and support policies and also with the thrust of agricultural research and extension effort which have been largely oriented towards these two major crops.

The Punjab government some years ago referred the matter to a committee of experts headed by Dr S.S Johl, a leading agronomist, which strongly recommended agricultural diversification for ecological and economic reasons, with at least 20 per cent of the land currently under paddy and wheat being turned over to oilseeds, pulses, maize, fodder crops, fruit and vegetables. Wheat requires four critical irrigations while paddy needs 25. Therefore, if kharif pulses, oilseeds and other crops could replace paddy this would be ideal. Sunflower, an oilseed, has gained popularity as a rabi crop since 1991. Within a year its acreage had increased from 14,000 ha to 85,000 ha with yields up to 1.6–1.8 tonnes per ha and a better market value than wheat. Though sunflower requires as much or a little more water than wheat, it helps break the paddy-wheat cycle. Paddy straw is no longer burnt but is used as feedstock for making paper. Experiments are afoot to size up its economics and other parameters for generating power. But the time gap between the paddy harvest in late April and paddy sowing in June is insufficient for organic matter to grow and then be ploughed back as a soil conditioner. Haryana too has gone in for sunflower and over 100,000 ha were under this crop in 1992. This could perhaps be combined with encouraging apiaries by small and landless farmers as bee-keeping is very profitable and sunflowers and floriculture require bees for pollination.

There is no call entirely to eliminate rice cultivation from Punjab and Haryana. Eastern India and the coastal plains must increasingly supply the rice the country requires, permitting Punjab, Haryana and Western Uttar Pradesh to diversify into other crops. An expansion of paddy acreage in Punjab and Haryana under superfine basmati varieties which command high prices and a flourishing export market would provide higher value on a smaller acreage while releasing land for other crop rotations.

This is the broad direction in which these States must move. Change is evident. Production of oilseeds, pulses and milk is rising. The Punjab experiment with foreign collaboration in growing tomatoes and potatoes with high-yielding seed and a package of improved practices has shown dramatic results. Some 4000 ha under tomatoes in Hoshiarpur district have yielded farmers incomes of up to Rs 60,000 per ha with less water use than the Rs 5,000 per ha which they used to earn previously. (Jha, July 1993)

Punjab's future lies in diversification, agro-processing and industrialisation. Fisheries are coming up, not only in brackish waters but in other water bodies. The largest of these are, of course, the Bhakra and Pong reservoirs where the fish catch was 855 metric tonnes and 485 metric tonnes respectively in 1991-92. Fishery cooperatives have been organised and the reservoirs are stocked by the Himachal government. The markets are in Calcutta, Delhi, Jammu and Amritsar. Fishermen's earnings have gone up to Rs 1000 a month but could easily double. Fishery development is a weak area of Indian water resource development. There is limited training and infrastructure, or cold storages in the Bhakra-Pong region and few if any refrigerated Railway vans. The completion of the Thein Dam in 1997 will create another large lake.

Promise of Hydel Energy

The commissioning of the Bhakra and Beas-Pong power systems gave Punjab and Haryana the electricity they needed to extend rural electrification to all their villages by the early 1970s. The spread of tubewells and the insistent demand for their energisation was certainly a compelling factor. Both states are now poised for an industrial leap forward with agro-processing (cotton, textiles, sugar, paper and strawboard, dairying, food processing) and light engineering in the lead. Future progress probably rests more on cheap and plentiful energy than on more water for irrigation.

Bhakra-Pong are basically irrigation projects and are operated to maximise agricultural production and meet farmers'demands. Yet, all their releases pass through turbines to generate a very substantial quantity of power. Sceptics initially wondered where Bhakra power was going to be used and 164 MW of Bhakra capacity was in fact committed at a rate of nine pies per unit to the Nangal Fertiliser Plant, using electrolysis technology to manufacture nitrogen. Before long, the demand for power began to grow by leaps and bounds and there came a time when farmers, once glad to get Nangal fertiliser, thereatened to damage the plant unless more power was released to enable them to operate their tubewells. They argued that without water, mere application of fertiliser would burn the soil. Nangal Fertiliser Ltd. has since adopted another technology to reduce its power demand.

The BBMB has improved the thermal-hydel mix in the northern grid and has made a notable contribution towards regulating frequency and voltage to stabilise the system which must operate within given parameters or else give rise to disturbances. The BBMB has performed well but faces unavoidable hurdles. Work on a 313-km 400 KV line from Dehar to Bhiwani was started in 1980 but was still incomplete nine years later on grounds of forest clearance, demands for additional compensation and delays by work-charged staff who feared retrenchment once the job was completed. Environmental clearance was given for the right of way in Himachal and Haryana but not for the Punjab portion. Compensatory afforestation was sought and was completed in the Ganguwal area. Yet, two stretches of 600 and 200 yards remained to be cleared. 23,000 trees were lopped, not cut, along the entire right of way and compensation paid for reconstruction of such houses as had been acquired. And this when the BBMB is only a project manager and generates power at 2.3 P per unit. The power is allocated to the partner States in agreed proportions with a 12.5 per cent royalty in free power going to Himachal as owner of the dam and power sites. The States then reimburse the BBMB in a given ratio. But some State Electricity Boards have been in arrears.

The BBMB system generated 13,000 million units of energy in 1991-92. Of this 50 per cent goes for agricultural uses, 20 per cent for domestic and municipal uses and the balance of 30 per cent to industry. Demand is growing and Haryana fears that by 1997 it will face a peaking deficit of 900-1100 MW. With more thermal plants coming up, the hydro-thermal mix is likely to become more adverse, as against an optimal 40 : 60 ratio. This will in turn affect the profitability of the Electricity Boards and entail far higher electricity charges in view of the higher costs of thermal generation.

There is a fall between Bhakra and Nagal of 30 to 40 feet which could be tapped to generate 75 MW at a cost of Rs 120–150 crores. This is under study.

The Thein Dam will add 600 MW of installed capacity by 1997. Beyond this, the Chenab, Ravi, Beas, Sutlej and Yamuna basins in Himachal have an assessed economic potential of 12,500 MW and J & K another 10,000 MW of which only a modest part has been harnessed. This represents a huge asset which could be turned to good account in generating wealth for the hills and turning the wheels of industry in the plains. Himachal and J & K have the sites but lack the resources, while Punjab and Haryana have the need. The BBMB has the expertise to develop and manage some of these resources if its charter is suitably expanded. Its design office, formerly in Delhi, was moved to Nangal in 1962 where it has since languished for lack of work. This is a pity for it could, like the Snowy River Authority of Australia, use its background and personnel to become a power consultant and design agency and enabled even to raise funds and construct mini and micro hydel plants in the neighbouring states. It could certainly be authorised to construct the 600 MW Kol Dam above Dehar and could in fact have done Naptha-Jhakri (1550–2050 MW) a little above the Kol site, which has now been entrusted to a new corporation. Basin planning and watershed management are closely integrated. The dispersion of the Bhakra, Beas and Ravi projects has a lesson. Consolidation rather than fragmentation is indicated.

The DVC was a good idea, modelled on the highly successful Tennessee Valley Authority in the United States. It was unfortunately not allowed to function for long in the manner conceived as a truly basin authority linking upper catchments and commands. The BBMB could assume such a role to the advantage of the concerned northwestern States.

Tourism is another aspect of the Bhakra-Beas projects that has not been adequately exploited. Militancy in Punjab from 1980 onwards did cast a shadow on free and easy movement. But that phase is hopefully over and there is a significant potential here that could be developed with advantage.

The BBMB's terms of reference merit review. While at present it is limited to managing existing assets, it is required to manage conflicts between the demands for irrigation and power, between the partner States, and between them and the Centre. It comes under the Department of Power at the Centre though its first charge is irrigation and, with it, drinking water to Delhi and Chandigarh. Though nominally vested with the powers of a state government, water and power rates are fixed elsewhere and it receives funds for operation and maintenance from the concerned states. These rates entail subsidies which, if economically fixed, could enable the BBMB to improve maintenance, develop its infrastructure, undertake new developments and provide a greater range of services than it is permitted to do under its present charter.

A Balance Sheet

Is then the Bhakra-Beas complex a disaster or looming castastrophe? The facts speak loud and clear. The project has been an outstanding success. It has built India and has long paid for itself many times over. The thesis that Bhakra and the green revolution it sustains has devastated the land with waterlogging, salinity and chemical toxicity leading to "soil death" is patently absurd. There are certain problems of land and water management but these are being addressed. Alarms about salinity and waterlogging have been sounded in India since the development of the Ganga canals 140 years ago. Not that this should breed complacency; but there is little cause for despair. There is no case whatsoever for abandoning or retreating from further development of available irrigation potential in the country with prudence and appropriate cropping and land use strategies. The test would be to ask the farmers of Punjab and Haryana what they think of Bhakra and whether they would be better off without it. The very question would be treated with derision.

The further notion that the Bhakra-Beas-led green revolution must inevitably have led to violence and in fact was a primary cause of terrorism and separatism in Punjab can only be attributed to the ultimate political and social innocence of the blissfully uninformed. It represents a total misreading of the development and context of events in Punjab, which in some ways even predate partition, and an astonishing lack of understanding of contemporary India. Of course economic and social discontents arising from a variety of circumstances will be grist to the mills of those who wish to stir up trouble anywhere and at any time. The very process of modernisation upsets traditional mores and balances. The trigger could be provided by roads, electricity, schooling, the eradication of malaria, radio and television, physical mobility, or a hundred other things, but most often by a combination of these. If the green revolution was the primary or even the most potent cause of conflict, then why did not terrorism and separatism spread to Haryana and why did Naxalite violence visit West Bengal in the mid-1960s or Telengana explode in the early 1950s?

The related nostalgia for a subsistence peasant economy, the Persian wheel and the rest by Vandana Shiva for example is again an incomprehensible rejection of modern scientific advance — not of crass consumerism and greed — in favour of some kind of sublimated primitivism which romanticises not simplicity but poverty and its accompanying degradation. The entire thesis ignores the tremendous impact of population growth which is again laughed off in some quarters as an elitist plot against "the people". Organic farming is all right but it is at present to the green revolution what the *charkha* is to the modern textile and garment industry — a supplement, not a substitute. If it heralds a real breakthrough, there should be every reason to adopt it more widely. But until then someone must take responsibility for feeding and clothing India and doing the thousand and one things that would ensure a better quality of life with equity, which is what governance and development is or should be about. It is for this same reason that blind opposition to biotechnology and the so-called "second green revolution" is so misplaced.

The Bhakra-Beas projects have provisioned the national larder and brought agricultural stability to Punjab and Haryana. In the bad drought year of 1987, when many parts of the country reeled under distress, farm production in northwest India was unaffected. Thanks to that, India could face and overcome the crisis.

Employment and income have risen appreciably and Punjab has attracted migrant labour for sowing and harvesting from as far afield as Bihar and Orissa. The pattern of employment will change significantly again with the "second green revolution", and will demand greater education and new skills.

A requiem for the dinosaur is not on the national agenda. And India minus Bhakra would not be a better place.

REFERENCES

1. Gulati, N.D., *Indus Waters Treaty: An Exercise in International Mediation*. Allied Publishers, New Delhi. 1973.

2. Bhakra-Beas Management Board, *History of Bhakra-Nangal Project*. BBMB, Chandigarh. October 1988.

3. Anon. Water Resource Projects: Rehabilitation and Resettlement of Displaced Persons — Norms and Measures Adopted. Bhakra Dam Project. Symposium on Large Dams, Small Dams: Socio-Environmental and Techno-economic Assessment. New Delhi, December 2, 1991.

4. Bhakra-Beas Management Board. *Annual Administration Report, 1989-90*. Chandigarh, March 1990

5. Shiva, Vandana, The Violence of the Green Revolution: Third World Agriculture, *Ecology and Politics*. Third World Network , Penang, Malaysia. 1991.

6. BBMB, "BBMB Releases from Bhakra Dam and Pong Dam". Note given to author. BBMB, Chandigarh, 1992.

7. Sahgal, Bittu, Unquiet Spring: The Root Cause. *The Sunday Observer*, New Delhi, July 18, 1993.

8. Mansata, Bharat, How Green Was My Revolution. *The Sunday Observer*, New Delhi, June 21,1992.

9. Sengupta, Amit, "How Green Was The Green Revolution?", A review of the film *The Good Earth* by Krishnendu Bose. *The Pioneer*, New Delhi, July 8, 1993.

10. Vander Velde, Edward J and Kijne, Jacob W., Salinity and Irrigation Operations in Punjab, Pakistan: Are There Management Options? Workshop on India-International Irrigation Management Institute Collaborative Research in Irrigation Management. New Delhi, February 13-14, 1992.

11. Jha, P.S., Punjab After the Terror. *The Hindu*, New Delhi, July 14, 1993.

12. National Land Use and Conservation Board, Report of the Committee of Experts on the Draft Outline of National Land Use Policy. New Delhi, 1988.

3

A Desert Brought to Life

THE Rajasthan Canal or Indira Gandhi Nahar as it was renamed in 1984 has all the makings of being to Rajasthan what the Nile is to Egypt, a ribbon of green parting the Great Thar Desert, restoring life to a once-verdant region. As Chief Engineer of the then Bikaner State just after Independnce, Kanwar Sain envisaged the Harike barrage on the Sutlej to meet the threat of Pakistan trying to outflank the key Ferozepur Headworks a little downstream, along the new international border. That sparked the idea of the Rajasthan Canal.

The preliminary project report on Harike he submitted to the Government of India in June 1948 concluded with the vision of irrigating "more than seven million acres of land in the Bikaner and Jaisalmer States . . . a fantastic dream . . . (that) will come true in less than a quarter of a century". He saw the Bhakra dam and similar Beas and Chenab dams as integral to this grand design. (Sain, Kanwar, 1978)

That hope is no more a mirage. The Canal has rounded Jaisalmer and is being hewn through dunes and arid, sand-blown wastes along the desert marches towards Gadra Road, about 1500 km from the Himalayan storages and upstream diversions that feed it. The historian, Toynbee, said of it that the potential fertility of Rajasthan's soil "needs the magic touch of water to bring it to life; and irrigation requires human daring . . . . This Canal is . . . (an) admirable adventure". (Toynbee, 1961)

"A Region of Death"

The Thar desert once enjoyed a less harsh environment with the lost Saraswati, the ephemeral Ghaggar and the wayward Yamuna flowing through it down to the Rann of Kachchh. The pre-Harappans settled the Ghaggar plain and, at Kalibangan, archaeologists have discovered a ploughed field with furrows following a grid pattern that permitted the sowing of two different crops, one of them barley. Radio-carbon dating places this find between 2300–2100 B.C. (Misra, V.N., 1977). The subsequent Harappan civilisation spread from Ropar in Punjab to the Narmada estuary and Kachchh. Farmers of that time are known to have used gypsum to fertilise the soil. (Savithri, R. 1976) The Harappan culture lasted till at least 1800 B.C.

Arid conditions are believed to have set in from anywhere around 4000 B.C onwards, imperceptibly at first, with possible orogenic changes influencing the river system which in turn resulted in falling groundwater levels. Xeric conditions were, as later, aggravated by the growth of both human and livestock populations, the extension of cultivation to marginal lands, and the loss of vegetal cover. Latterly, rodents and locusts have contributed to desertification. Maybe all of this induced climate change and groundwater salinity as precipitation declined. Although western Rajasthan nowadays receives little rainfall, atmospheric humidity is comparable with that over Bombay, Trivandrum and Nagpur. The area is exceedingly dusty, with particles being blown up to 20,000–30,000 feet. It is surmised that "the dust over this arid zone is the main cause of the mid-tropospheric divergence that occurs over the region which does not allow sufficient precipitation to occur in spite of good atmospheric humidity". (Krishnan, A. 1977)

This is an interesting observation. For, with the Canal thrusting into the desert, two things will happen. Farming and vegetal cover will enhance evapotranspiration, while sand dune reclamation and stabilisation, crop cover and extensive afforestation and the shelterbelts now under plantation will reduce dust. Whether the combined effect on the micro-climate could in time enhance precipitation over the region remains to be seen. This appears worthy of study.

The desert is in motion, with winds blowing southwest to northeast. When work stops or slows down elsewhere in the country on account of heavy rains and floods between June and September, Rajasthan's dunes start rolling like waves, engulfing new areas, choking farms and canals, roads and rail tracks. High winds erode the topsoil leaving behind hard gravel. Dust clouds the atmosphere while heavier sand particles swirl and eddy at ground level, with columns pirouetting like ecstatic prima ballerinas, skirts flared, gliding across the desert stage. Ludarwa, an earlier seat of Jaisalmer state, lies buried in the sand like so much else. Only the temple, now renovated, pierces the surface. Raja Jaisal constructed the present city around 1156. Tod describes the Great Indian Desert as Maroosthali or Region of Death. (Tod, James. 1832)

The Aravalli Range separates the Indus and Ganga drainages. Discontinuities in this chain of hills mark ten gaps or windows through which the desert sands have been propelled northeast to cover parts of Ajmer, Nagpur, Jaipur and Sikar districts, as evidenced by satellite imagaries taken between 1972 and 1983. (Anon. Aravalli 2001 A.D Conference, 1987) The Indira Gandhi Nahar Pariyojana (IGNP) will anchor the desert.

Indus Waters Treaty

Not all these possibilities were necessarily part of the early vision of the Rajasthan Canal. But irrigating the desert to grow more food was a prime consideration, especially in view of the transformation wrought immediately to the north through the Gang Canal, of which this would be an extension. Kanwar Sain records an interesting proposal for an ad hoc settlement of the Indo-Pakistan canal water dispute mooted at a meeting in Karachi in March 1951. The Pakistan negotiators, Sain writes, were willing to let India have all the Ravi and Beas waters being utilised in Pakistan canals prior to partition. The Indian negotiators were agreeable provided India had the right to divert such waters of the Chenab as it could tap at Marhu in East Punjab. Any deficiency in Pakistan could be made good by the construction of a (2.5 MAF) storage dam on the Chenab in Indian territory at Dhaingarh, J & K. Pakistan was unwilling to concede India the right to any storage on the Chenab and the matter was dropped. However, Indian engineers on subsequent investigation found the site infeasible for lack of suitable foundations.

Discussion on the Indus waters dispute continued under the good offices of the World Bank. Pakistan's "preliminary outline plan" submitted to the Bank entailed a demand for 506 MAF of water, or over three times the total supply of the Indus system calculated at 168 MAF. The scheme defined all undeveloped areas in India as lying outside the basin (Sain). The Indian outline plan envisaged storages on the Indus, Chenab, Ravi, Beas, Sutlej and a tributary of the Jhelum, aggregating 17.7 MAF in all, enabling India ultimately to use 37 MAF of water, including 8.1 MAF already utilised, and Pakistan 83 MAF including 63.4 MAF already utilised. Neither proposal made much headway.

In April 1950, the Rajasthan Government was instructed to survey the areas that might be irrigated by a canal taking off from Harike. By 1954 the Government of India had in consultation with the concerned States determined the allocation of Indus waters between J &-K, (East) Punjab, PEPSU (the Patiala and East Punjab States Union) and Rajasthan under the World Bank's scheme.

This would allocate the entire waters of the three eastern rivers, namely, the Ravi, Beas and Sutlej totalling some 33 MAF to India. Rajasthan's allotment was 8 MAF from the Ravi and Beas with storages contemplated on these rivers at Thein (currently under construction) and Pong (completed in 1976).

The Pong Dam project was administratively approved in August 1960 but it had been hoped to divert waters into the IG Nahar to irrigate up to 900,000 acres under Phase I by 1962 from the tail of the Bhakra and Sirhind canal systems. Progress was slow and only 90,000 acres were brought under irrigation by 1965-66 and 450,000 acres in 1971-72. By then India was entitled to withdraw the entire waters of the three eastern rivers for its own uses on the expiry of a ten-year transitional period after the signing of the Indus Waters Treaty by when Pakistan was to complete its replacement works for waters earlier withdrawn from these rivers.

The Rajasthan Canal was sanctioned in July 1957 and was to be completed by 1965. The delay in the project saddened the country's prime Indus Waters negotiator. Wrote Gulati: "The Rajasthan Canal, which was the kingpin of India's case in the 1950s for additional uses from the Indus rivers, which was the *raison d'etre* of withdrawing the Eastern rivers waters from Pakistan, and the principal purpose underlying several successful battles fought by my colleagues and me, a development which the head of the Government of India, the accredited leader of the

Indian people for decades (Nehru) had declared in a formal communication to the President of the World Bank in 1960 as "of vital importance to us" is, 12 years later, still a project in the early stages of execution". With 90 per cent of the CCA still to be developed, the picture presented was different from "new irrigation colonies emerging in Rajasthan in lieu of Lyallpur and Montgomery (which) we had lost to Pakistan". (Gulati, N.D., 1973)

Pakistan had initially strenuously opposed the very concept of irrigating the Western Rajasthan desert. A first feasibility report was prepared in 1953 and accepted by the World Bank team after an aerial reconnaissance. Gobind Ballabh Pant turned the first sod to inaugurate work on the 204-km Rajasthan Feeder in March 1958. Shortages of coal and cement delayed its completion. Yet water was released into the first distributary, Noraog Desar, by the Vice-President, Dr. S. Radhakrishnan, through Bhakra channels in October 1961. This background is essential for an understanding of the Rajasthan Canal and offers a clue to certain later developments.

Complex Features

Rajasthan's final share in the allocation of Indus waters was in 1981 increased to 8.60 MAF (10.6 bn cu m) of which 7.59 MAF was allocated to the IG Nahar, the balance being apportioned to service the Gang canal and Bhakra system within the State. The waters of the Ravi, which are to be stored behind the Thein Dam near the J & K border, are diverted into the Beas through a link canal at Madhopur a little before the river enters Pakistan. The Beas in turn is partly stored behind the Pong Dam near Talwara in Hoshiarpur district, with its 8 MAF reservoir backing into Himachal, like Bhakra's Gobindsagar lake. The river is also partly diverted further upstream at Pandoh to fall into the Sutlej after generating power on a drop at Dehar. These waters then flow into the Gobindsagar reservior and then pass through the Bhakra turbines to commence the long journey to Harike, several hundred kilometres away, there to be diverted into the Rajasthan Feeder.

The Rajasthan Feeder continues as the IG Main Canal extending 649 km in two stages to Mohangarh. The Sagarmal Gopa

Branch takes off here and will run another 300 km to tail at Gadra Road near Barmer making this a longer waterway than either the parent Ravi or Beas rivers. This is truly a herioc concept and already a considerable reality.

The Indira Gandhi Nahar is not merely a great feat of Indian engineering but an act of faith. The project is intended to irrigate 1.388 m ha of land on full development, 301,000 ha of this through six lifts and the rest by flow. Of the total water allocation, 11.5 per cent or 1200 cusecs year-round is committed to providing municipal and industrial water for a huge thirsty population.

Several features of the command area are noteworthy. Much of it, especially in the lower Stage II areas, is very sparsely settled and supports an essentially nomadic, pastoral population. Rainfall is scanty, irregular and uncertain. Much of the groundwater is saline. The Ghaggar loses itself in the desert but when in flood reaches Hanumangarh and sometimes spills into Pakistan. Further south, there is no natural drainage outlet. The land slopes from northeast to southwest, towards Pakistan, with sand dunes up to 60 m high interspersed with inter-dunal flats and a large plain in the middle. There are a number of depressions along the Ghaggar and further southwest, a series of ranns or mud flats forming a giant stairway down to the Great and Little Ranns of Kachchh. The desert experiences extremes of climate and supports a sparse vegetation of sturdy khejari, Accacia tortullus, Rohida (a desert teak), Akra, a succulent bush, and nutritious sewan grass. This cover has been lopped, cut and over-grazed to the point of regression over wide areas, leaving behind unpalatable scrub. Some orans or sacred groves remain and where the root stock has survived, regeneration quickly follows closure. The Bishnois have come to be regarded as keepers of the desert. There is a scattering of small so-called perched aquifers and a few deeper aquifers, notably the Lathi and Chandan — Dabla series near Pokharan where the 1974 nuclear test implosion site attracts little curiosity. Camel, cattle, sheep and goat are the pastoralists' wealth, and herds of lithe Chinkara dart through the desert. In drought years, whole villages have habitually migrated north towards Punjab or east to Mewar's greener pastures. Traditional water harvesting techniques were ingeniously devised over the centuries and moisture conserved in khadins, bunded tanks, or stored in family kunds carefully conserved under lock and key. Nevertheless, everything shrivels and dries when the rains fail, as they often do, sometimes for years together. Some 84 deserted Paliwal villages around Jaisalmer stand mute testimony to a once-flourishing khadin cultivation. Hurtful taxation by a greedy chieftain whose roving eye fell on a Paliwal belle is said to have caused the clan to migrate almost overnight *en masse* to Vidharba about 175 years ago.

Stages I and II

This is the land to which the IG Nahar has come. The canal slices the desert. Migration has been reversed; people and livestock now migrate to its life-giving waters. A unique feature of the Project is that it entailed no displacement as there was virtually none to displace. On the other hand, and more especially in Stage II, irrigation requires colonisation. Ultimately, some two million people will have settled in the IG command to create virtually another country.

Few thought the Canal would ever materialise. There were tears of joy when the waters were first released at the head in 1961. For many years thereafter, those nearer the tail towards Jaisalmer gave little credence to the prospect of getting water. How would a canal come through the desolate desert? And who would let the waters flow, as rich farmers and thirsty villagers higher up the system would use it all? In 1987 the waters reached Mohangarh at the terminus of the Main Canal. Now people further along towards Gadra Road, in Barmer, wonder whether the Canal can or will ever reach them. It will by 2004 at the present rate of progress, but could be completed by 2000 if more funds are provided.

The Project is being constructed in two stages. Stage I is complete. It consists of 189 km of Main Canal, several branches and distributaries, and is irrigating 528,000 ha at 110 per cent intensity. Village abadis, agro-service centres and market towns have been established. Water courses have been lined, shelter belts planted, roads and water supply 'diggies' constructed, and other amenities for education, health, and so on created under the command area development and colonisation programmes. With an investment of Rs 540 crores on Stage I, the annual agricultural output in 1991-92 was valued at Rs 550 crores, yields of cotton and groundnut ranking among the highest in the country.

Stage II entails extension of the Main Canal by a further 256 km. This was completed in January 1987. Work on several branches, distributaries and lesser order canals are under construction, with related colonisation and command area development works. Some Rs 2700 crores had been spent on Stage II until March 1992 and another Rs 5420 crores may be required (with an assumed eight per cent price escalation per annum) to complete the project by 2006-07. The IG Nahar Board is, however, confident of foreshortening this period by up to six years if funds are made available. Stage II will provide irrigation to an additional area of 940,500 ha and yield agricultural produce worth over Rs 1000 crores per annum. With other benefits, this is expected to give an economic return of 18.84 per cent.

Stage II is not merely larger than Stage I but significantly more difficult as the Canal enters a more hostile desert and dune environment requiring extensive dune stabilisation and the development of civic and economic infrastructure. Incidentally, the Project includes six lift schemes, five of these in Stage II, to irrigate 347,000 ha in all and provide drinking water supplies at the rate of 50 and 130 litres per day to a number of cities and rural areas. The lift is being limited to 60 metres, except for the Johdpur canal which will entail a lift of 145 metres, and leads of up to 200 km. The benefit zone will radiate far beyond the irrigated command and will cover parts of Bikaner, Jaisalmer, Barmer, Jodhpur, Churu and Sri Ganganagar districts.

Skepticism about the IG Nahar Project has been dispelled in many minds. But there are those who continue to criticise it on other counts, especially the likelihood of mounting waterlogging and salinity and its failure to uphold equity in the matter of allotments and grant of khatedari rights. The fate of the Pong Dam oustees, who were to be resettled in the IG Nahar command, deteriorating health conditions, corruption and a seeming decline in the status of women are among other negative factors mentioned. These need to be answered.

Waterlogging and Salinity

There has undoubtedly been some waterlogging and salinity,

particularly in Stage I. Much of this, and the worst affected by far, has been by seepage from water diverted and stored in a series of Ghaggar depressions on the left side of the Canal. Other areas are liable to be affected by the existence of a hard pan strata underlying stretches of the command area, principally in Stage II. A hard pan is defined as a layer having a porosity of one-fifth or less of the surface strata and is therefore not totally impervious but retards infiltration. It has been classified in three categories of 0–10 metres below ground level, 11–30 m bgl, and more than 30 m bgl. Cropping patterns and irrigation intensities are being recommended accordingly. An Operations Research Group (ORG) survey cites 33 and 76 per cent of the flow and lift commands as having a hard pan layer 0–10 m bgl and prone to waterlogging if adequate remedial measures are not undertaken. (Ramanathan and Rathore, 1991)

Having monitored 800,000 ha in Stage I of the IGNP and parts of the Gang Canal and Bhakra commands in Rajasthan which have been under irrigation for 60 and 30 years respectively, Water Power Consultancy Services Ltd has found the area actually waterlogged in this region to be 8600 ha. Another 17,000 ha is critically affected with the water table within 1.5 m and a larger area of 198,000 ha is "potentially sensitive" with the water table being within six metres of ground level. It believes that waterlogging and salinity hazards might manifest themselves sooner in Stage II on account of the existence of hard pan layers at shallower depths. (WAPCOS, 1993) About 1000 ha of Stage II is waterlogged, mostly as a result of canal seepages.

The World Bank had opined that the IGNP command lacks irrigable soils over significant areas and irrigation could lead to waterlogging and salinity in the absence of proper drainage. Earlier, FAO had done an elaborate Stage I soil survey and strongly advised against running the canals in the Anupgarh region which it found unsuitable for irrigation. It advocated extensive drainage and sprinkler and drip irrigation.The IGNP authorities differed and went ahead. Anupgarh is now a garden.

The IGN Board is not unmindful of these cautions and has paid due heed to the experience in Stage I. But its experts believe that the hard pan has needlessly been made a bogey. It has analysed the factors that led to waterlogging in Stage I. A number of planning and monitoring studies have yielded what is seen as an appropriate strategy for Stage II alongside plans for reclaiming lands affected by waterlogging and salinisation in earlier phases.

Delays in the project, largely on account of a paucity of funds, have resulted in quantities of water continuing to flow down the Sutlej below Harike to Pakistan. This is something the Government of Rajasthan and the Central Government have been concerned to limit if not altogether avoid for fear that sustained flows to Pakistan could lead to fresh claims regarding established rights and prior appropriation despite the clear allocation of all the waters of the Eastern rivers to India under the Indus Treaty. Rajasthan, for its part, has been equally anxious to protect its share of waters in the inter-State allocation of Ravi-Beas surpluses between Punjab, Haryana and itself. Covetous eyes have also been cast on Rajasthan's unutilised share by the upper riparians.

Various other factors are responsible for this situation. The Thein Dam will not be ready to store Ravi flood waters until 1996-97. Stage II of the IGNP command is still drawing only limited quantities of water as not only is the irrigation network incomplete but lands have to be allotted and the newly commanded areas colonised before utilisation is possible. In the absence of intermediate storages, fluctuations in demand, especially those caused by sudden rainfall, necessitate spillage down-river into Pakistan. As the dams on the Beas, Sutlej and Ravi are only designed to store mean year flows, surplus water in years of heavy rain must be allowed to escape. This also applies to run-off from the free catchments below these dams and from other unregulated minor catchments in the hills. The requirements of power generation do not also have an exact fit with irrigation needs despite monthly meetings of the Technical Committee of the Bhakra-Beas Management Board. Day-to-day releases cannot therefore be fine-tuned beyond a point and surpluses must run to Pakistan. A detailed study of surplussing at Harike from 1979 to 1989 shows that water has had to be allowed to escape even during the 1987 drought, let alone the following year when there were record floods, as again in 1993. Taking account of the holding capacity of the IGNP canals, including the

Rajasthan Feeder, and the possibility of diverting surpluses into certain depressions within the command, it is estimated that 0.3 MAF can be stored. Allowing for evaporation and seepage losses, this may leave a net 0.22 MAF available for further utilisation.

The Pakistan factor has been a constant worry. This led at the very outset to a decision to provide for a delta or water-use allocation in the IGNP Stage I command in excess of that provided to the adjacent Gang and Bhakra commands which, being in the Sutlej valley, have loamy soils with better irrigability. The figures are revealing. Between 1981 and 1989, the average rate of water released at the head of the Feeder Canal was around 1500 mm as against the designed value of 723 mm. The corresponding values in the neighbouring Gang and Bhakra commands during 1988-91 were 526 and 462 respectively. To this must be added 0.58 MAF or the 715 mm annual inflow into the head reaches of Stage I, all of this contributing to groundwater recharge. (IGN Board, 1993) The engineers and planners who were involved in the early phases of the IGNP brought the experience of the Bhakra and Gang canals with them. In view of the more difficult desert terrain encountered, direct outlets were allowed from the Main Canal at the rate of 10.46 cusecs per 1000 acres in order to stabilise irrigation. While this practice may have had an initial logic it was obviously unsuited for long term usage. A water allowance of 5.25 cusecs per 1000 acres was allowed in Stage I more generally. Since the lower reaches of the command were not yet ready to draw water, with Stage II still further beyond the horizon, farmers in the head reaches departed from the planned cropping pattern to grow more water-intensive crops and, in the manner of all head-end irrigators, diverted surplus waters into their fields in the common belief that more water ensures a better crop. In the first (Kanwar Sain) lift, power failures aggravated uncertainties in field deliveries, thus impelling farmers to maximise their use of water when available. In the absence of drainage or conjunctive use through tubewells, the water table kept rising.

A more acute problem arose in the effort to divert away from Pakistan and store the flood waters of the Ghaggar in 18 natural depressions running parallel to the left bank of the IG Nahar. With plenty of water in the canals and no system of utilising the stored water for irrigation beyond limited local uses, the stagnant water seeped into lower-lying farmlands to cause severe waterlogging as at Badopal. Fields have been submerged and houses eaten away, leaving behind saline residue. Draining the Ghaggar depressions was also thwarted by contractors who had bought fishing rights in these depressions by auction and were not inclined to lose a profitable business. The affected villagers have been assisted to rebuild their habitations on higher ground but have not been amenable to exchanging their lands with allotments in other areas. Some would appear to prefer to wait for reclamation works to restore their lands in view of the fact that these tracts have developed and are otherwise visibly prosperous. But it is also alleged that payment of compensation has been tardy, leaving those affected with little option but to remain.

Improved Water Management

It is now advocated that the 18 inter-connected Ghaggar depressions, which fill in sequence depending on the extent of rainfall, might be lined with low-density polyethelene (LDPE) so as to prevent seepage, and the net 0.22 MAF storage be used to irrigate an additional 100,000 ha. Consideration is being given to this proposal which would require an investment of Rs 400 crores. Its execution could certainly greatly ameliorate if not altogether reclaim the worst waterlogged areas.

A number of other measures are being planned. The water allowance for Stage I is being reduced from 5.25 to 3.25 cusecs. A beginning has been made in the Kanwar Sain lift command where the groundwater has been lowered by reducing the water allowance from 3.9 to 2.73 cusecs. The whole IGNP Command is to be monitored through a network of 500 piezometers and observation wells which is already quite advanced and soon to be completed. Computerised readings will enable project administrators to plan water management strategies appropriate to each sub-region within the command. Current readings have shown a water table rise averaging 0.80 m per annum which must obviously be regulated before any further damage is done.

Structural inadequacies such as provision of direct outlets from primary canals are being remedied. A parallel channel is being constructed along the Kanwar Sain Lift from which regulated outlets will be provided. This will also make for equity so that tail-enders in any sub-command do not suffer.

There has been a problem with saline aquifers which, however, have often been found to have a thin layer of sweet water overlaying them in Stage I. With irrigation recharge, the fresh water layer is growing and it is becoming possible to blend it with brackish water within the limits of crop tolerance. A number of tubewells and dug wells is being commissioned to provide vertical drainage. Some of these wells will pump water and put it back into the canals for redistribution.

Warabandi or rotational watering is being practised to ensure equity as between irrigators as well as the larger objective of maximising crop productivity per unit of water. However, warabandi has resulted in emphasis on irrigation per unit of time, despite formation of water user associations through chak samitis which have a consultative role. The effort now is to encourage more scientific and productive border strip and furrow irrigation to spread water more evenly among the plant population within each plot. The Command Area Development authorities are being persuaded to try new water-saving crop rotations and to change traditional watering practices. Thus three to four waterings for gram and four to five waterings for mustard have been reduced to two to three waterings. (Commissioner Area Development, 1991)

Stage I experience and mapping of the more extensive hard pan strata have resulted in a new strategy for Stage II. The intensity of irrigation is being reduced to 86 per cent as against a nominal 110 per cent in Stage I with actual utilisation going up to 120–125 per cent. The water allowance is being reduced by 1.5–3.00 cusecs and no land is to be allotted in areas where the hydro-geological barrier is up to 0.5 m below ground level. No canals or water courses will be constructed in areas where the hard pan is up to 5 m bgl. Farmers will have the option to lift water and use it sparingly or to take up sewan grass cultivation and agro-forestry and develop pastures in these areas.

Bio-drainge through plantation of eucalyptus and other water-loving trees has yielded good results. Shallow hard pan areas with perched water tables will be treated with an appropriate mix of land and water management practices and waterlogged areas, as in the Lunkaransar lift in Stage I, will be de-watered through sub-surface pipe drainage systems. With the water table over a large part of the Command varying from 30 to 60 m in depth, an annual potential evapotranspiration of between 1800 to 2500 mm and an annual application of 510 m of irrigation, it is proposed to monitor the groundwater regime and see how long it takes for aquifers to approach saturation and then plan a drainage system. Model studies are programmed with Canadian (CIDA) assistance. WAPCOS, the project consultant, has recommended 560 km of drains and 896 shallow tubewells along the Main Canal and a larger number of shallow dug wells to intercept and recycle seepage. Canal use is to be matched with land capability with emphasis on pasture and livestock development, horticulture and a revision of colonisation and settlement policies to fit. (IGN Board, 1993) Remote sensing and other technological aids will be employed. But efficient water management and greater beneficiary participation will be crucial.

Water must be priced to meet its economic cost. Like elsewhere in India, it is mistaken policy to underprice water. This results in over-use and waste. It disables the authorities concerned from meeting operation and maintenance charges, which in turn leads to deteriorating and uncertain service to complete the vicious circle. The present water rate in the IGNP command averages Rs 100 per ha while some farmers are known to sell their (warabandi) "surplus" at a rate equivalent to Rs 1500 per ha or to buy far more expensive tubewell water. The IGNP requires Rs 16 crores per annum currently for O & M but nets a revenue through water charges of only half that amount.

The idea of constructing a drainage ditch to the Rann of Kachchh has been mooted but would appear to be expensive and could entail lift. Apart from the various measures suggested, salinised areas can be leached with freshwater. Salinised run-off can be injected into deep aquifers or led into depressions or ranns near Jaisalmer which could be used as evaporation ponds or solar stills. Brackish water can be used to breed brackish-water fish or to extract salt as from natural formations. Interesting ideas but problematical.

Waterlogging and salinity therefore exist and pose a more widespread problem in Stage II given a business-as-usual scenario. But with proper monitoring, drainage, land use and water management, not only can further waterlogging and salinity be prevented but the currently affected area reclaimed. It is in fact proposed to reclaim a waterlogged area of 9260 ha by 2000 and take preventive measures to obviate the problem elsewhere. Use of more sophisticated water-conservation methods such as sprinkler and drip irrigation have been advocated. Much depends on the nature of the crop, the topography, wind speeds, availability of energy and other parameters. Drip irrigation is suited to high value crops. Israel has a great deal of experience in water conservation and irrigation management in arid regions and it is proposed to get some Israeli experts to study the situation in the IGNP and train local people to implement its recommendations.

Settlement Policy

The other major issue that has attracted attention pertains to equity. The IGNP is unique in that it virtually displaces nobody but will settle, directly and indirectly, well over a million people in the command which has a diminishing density going south. Large parts of the desert are not permanently settled but are home to nomadic pastoralists who move from place to place seasonally where there might be water. Colonisation policy was set out by the Rajasthan Colonisation Act of 1954 which accepted a murabba or 6.32 ha plot as an economic holding. Such plots were preferentially to be allotted to landless persons on an annual basis pending formulation of permanent allotment rules. Such temporary allotments in fact continued until 1975 with 30 per cent reservation for scheduled castes and tribes, the allottees being selected on the basis of staggered proximity to the settlement-allotment site.

Under the Rajasthan Colonisation (Allotment and Sale of Government Land in the Rajasthan Canal Area) Rules, 1975, temporary cultivators, ex-servicemen, agricultural graduates, Bhakra landless, and landless of the same or neighbouring village or tehsil are eligible for allotment of government land up to one murabba, subject to the allottee's total holding not exceeding the ceiling. This allocation pertains to land "commanded" by the IG Canal. But if "uncommand" land is allotted, that is land such as dunes or upland patches within the command but too high to be irrigated by the Canal, it shall be rated equivalent to half of command land. Allotments are made from among applicants through draw of lots, with given areas being reserved for agricultural graduates, ex-servicemen and other prescribed categories in consultation with an advisory committee. Until the end of March 1992, over 118,000 allotments has been made (nearly 92,000 in Stage I and the rest in Stage II), with about 32,000 of these being in favour of scheduled castes and tribes. By that same date some 2.17 million acres had been allotted, something under half of this being "uncommand" land which farmers would have to reclaim and level gradually with considerable effort and investment.

The Rules also provide for sale of notified lands by special allotment to the eligible categories or, where such persons are not forthcoming, to any other bonafide person resident in Rajasthan for not less than 10 years from the date of application, subject to the land ceiling prescribed in the State. With 25 bighas making a murabba of 6.32 ha, the reserve price fixed for normal allotments ranged from Rs 500 to Rs 800 per bigha for sandy loam, light loam and 'nali' lands and Rs 150 per bigha for 'uncommand' land. By 1990, and several revisions later, the reserve prices had gone up from Rs 1250 to Rs 2500 per bigha for command land and to Rs 350 for 'uncommand' land. The special allotment price was simultaneously fixed at between Rs 4000–6625 per bigha for command land and Rs 2000 for 'uncommand' land , the price of command land coming down progressively as one travelled down the Canal into newly settled areas.

The price of normal allotments is to be recovered in 15 instalments, with a year's extra grace for scheduled castes and tribes. In the case of special allotments, half the amount is to be paid before actual possessson of the land allotted and the balance in two instalments within a year.

The price of a murabba for normal and special allotments is a maximum Rs 62,500 and Rs 1,65,625 respectively whereas the market price currently would be at least Rs 10 lakhs.

Utilisation of the irrigation potential created in Stage II is lagging because of tardy allotments and settlement, delays in respect of the one adversely affecting the other and causing canals to be choked with sand or infested with weeds, and roads and other infrastructure to be covered with sand — all this entailing additional expenditure for clearance. Accordingly in 1983, the Government of Rajasthan decided to reserve half the culturable area for special allotments. But by March 1991 only 4259 ha or 2.5 per cent of the allotted area had gone to 606 special allottees. Again, of 24,480 normal allottees, a bare fifth had actually occupied their lands because of adverse living conditions in a harsh environment, the large ratio of 'uncommand'to command area that was allotted, non-availability of water at naka points in the absence of lined watercourses, lack of infrastructure, choking of lower order canals by sand, and poor maintenance of chak infrastructure by some settlers. (WAPCOS,1993)

The 'uncommand' areas require considerable land-shaping which calls for skill and experience. Lined watercourses have encouraged settlement but by and large the rate of settlement has been low. The fragmentation of holdings between command and 'uncommand' areas and the distance between the allotted land and the abadi site have also operated as disincentives. When in order to overcome this some allottees have built farmhouses or dhanis on their plots, this has discouraged development of the abadi and related civic amenities. Random allotments have also resulted in communities drawn from the same village of origin being scattered, thus losing out on the kind of support community networks would have provided. Distance from the mandi and 'diggi' (drinking water source) and the quantum of irrigated land allotted have been factors affecting the rate of settlement. Non-availability of tractors for land-shaping, or inadequate credit facilities have also led to absentee landlords who have been content to let out their lands to sharecroppers who have been reluctant to make capital investments. Villagers on the left bank of the Main Canal, which lies outside the command area, have also become sharecroppers of absentee landlords within the command.

Another problem arises from the Rule that an allottee must start repayment as soon as any plot in his chak is irrigated, even though not his own. Tail-enders are severely disadvantaged.

A number of suggestions have been made from time to time to improve the rate of settlement. Free rations for settler families for two years and a cash allowance have been given under the World Food Programme linked to the construction of watercourses in undulating plots with dunes. Motivational officers have been appointed and motivation camps established. Efforts are being made to provide more infrastructure in the settlements. Changes in the schedule of repayment and levy of penal interest have also been recommended. Better coordination and synchronisation of efforts by the IGNP, the Colonisation Commissioner's department and the command area development authorities is obviously necessary. These are separate entities. One missing element would appear to be the absence of non-governmental organisations which could play an intermediary role. They could be actively encouraged. Better information services could also be provided.

The pattern of abadi settlements is perhaps too dispersed to instil confidence and a sense of community among allottees. An abadi for three instead of four to six chaks has been suggested. If, despite improvements and reforms, allottees are still not prepared to settle or, as happens, remain absentee landlords and let out their lands to sharecroppers then, it is argued, allotments should be cancelled. Those who prefer to wait to settle until the land is fully developed, impede the very process of land development, utilisation of irrigation and settlement. It is certainly true that whereas settlement was not a problem in Stage I it has become so in Stage II on account of more hostile natural conditions, greater dune formation, the sheer desolation and remoteness of the area with extremes of climate ranging from 48 degrees Celsius in the summer, with fierce sandstorms appreciably increasing the 'wind-heat' factor, to freezing temperatures in winter. Since the land has to be levelled and soil conditions gradually improved through cultivation, manuring and fertilisation, early returns are modest and might even entail a short term loss of income in relation to previous earnings at the last place of residence. This could be aggravated by incurring debt, anxieties over instalment repayments and threats of penal interest on default. If, however, the Government has constructed and maintained the water courses for years after settlement then, it is felt, cancellation of the allotment would be justified.

Equity and Efficiency

There is, however, another problem that has not yet been fully resolved. Not all the command in Stage II is made up of Government land. With the enforcement of the Rajasthan Tenancy Act in October 1955, khatedari (ownership) rights were vested in 1.2 million farmers who had any kind of relationship with a jagirdar (Section 15). However, three years later, a new Section 15A took back all lands so conferred. These became Government lands. Following an agitation in 1969, all tenants possessing lands prior to October 15, 1955 were given one murabba of land free and entitled to purchase additional land up to the prescribed 48 bigha ceiling (if command land). The Supreme Court ruled that the 1955 cut-off date was arbitrary. The Government introduced some modifications but the High Court struck down Section 15A as ultra vires as no compensation had been paid in terms of the then Article 31-A, a fundamental right.

In 1979, Section 15 AAA was introduced and revised in 1983 conferring khatedari rights on all pre-1955 farmers who sent in the requisite applications. Further refinements were introduced in 1984. Nevertheless, the position in Stage II is that many farmers have not yet received khatedari rights. This not only handicaps them in securing credit in excess of Rs 25,000 as may be required for purchase of tractors, but also engenders a sense of insecurity about investment in cultivation in view of cases of double allotment and land grab. (URMUL Trust, 1992)

The official response is that farmers have secured khatedari rights even without proof of title to land. They have been allotted one murabba free and enjoy the option to purchase additional land up to the ceiling limit at the reserved price. The Command being along the border, the allottees' antecedents have sometimes to be verified. Khatedari rights are also sought to be given after the allottee has repaid the land cost in the stipulated 16 instalments. However, if the allottee is not allowed to sell the land until after he has acquired khatedari rights, benami transactions are encouraged, leading to corruption which is alleged to be widespread among lower officials such as patwaris.

There has been criticism in certain quarters of the policy of special allotments, even to non-farmers, which has resulted in land speculation. With the best lands allegedly being reserved for special allotment, the charge is that the local landless have been marginalised and allotted the poorest lands which are therefore the most difficult to reclaim, resulting in discrimination, debt and a sharp departure from the intended social policy that the IGNP was committed to pursue. Rather than settle marginal farmers on marginal land, the emphasis should have been on inviting "entrepreneurs" to take up the challenge of reclaiming marginal lands in remoter parts of the command. Even if the special allottee is a more successful farmer than the poor landless labourer, is social equity not as important and even an essential element of economic success? In the fundamental conflict between equity and efficiency, the poor must be made efficient and productive. This process might be encouraged were project planning and administration more participative. If people are made to feel they own the system, they will make it work and take responsibility for its proper maintenance. "The IGNP is for the people, not the State. But where are the people, other than in their role as passive recipients?" Instead of viewing the allottees as a community, the Government continues to treat them as individuals. It is easy to say apropos the waterlogging problem in Stage I that farmers should grow less water-intensive cash crops. But, burdened with debts incurred for reclamation and sustenance, the farmer has no choice but to grow the most remunerative crops. Likewise, old community sanctions, norms and structures have broken down. Even the way land is visualised varies greatly. For the poor settler it is a matter of identity, security and survival. For the special allottee it is a resource to be exploited. (URMUL, 1992)

These arguments cannot be dismissed. Social justice was consciously built into the planning of the IGNP and is indeed a basic pillar of the Constitution and a Directive Principle. Yet neither can the efficiency factor be ignored. Settlement in Stage I was successful not just because conditions were less difficult but because the allottees had more farm experience which they could bring to bear on the land. Making land grants in the more difficult Stage II to poor labourers lacking this experience when required in even greater measure, was certainly in keeping with valid social policy but proved counter-productive as much as it has largely failed elsewhere. There is an important lesson here. Poverty alleviation and equity are often served far more effectively by rapid development and growth rather than by distribution of a limited cake. If "efficiency" is assured, then "equity" could follow as there is more to go round. But the converse is not true. Equity in certain conditions, such as the IGNP Stage II, does not guarantee "efficiency". If "efficiency" is not realised, "equity" must remain an empty promise, a social waste and a national liability.

The very challenge of the IGNP, especially in Stage II, requires hardy pioneers with the will, experience and wherewithal to overcome the harsh environment, the remoteness of the region and a minimal gestation period of sweat and labour in order to succeed. It requires stamina and grit and a cheerful acceptance of hardship before reaping any reward. The "equity" approach ignored this factor whereas the "efficiency" factor would bank on it. Not that equity considerations need have been altogether set aside. The mix of "equity" and "efficiency" could have been and could still be different, with a sufficiency of "effficiency" built into the programme to carry a degree of "equity" with it.

Large investments are being made in the face of severe financial constraints which have already grievously slowed down the IGN Project. There is an opportunity cost which the nation unavoidably pays if such investments do not mature and start yielding the expected dividends soon. This burden must ultimately affect genuine "equity" far more than cosmetic "equity" based on political or social expediency which is soon exposed as hollow and unsustainable. The great game of "equity" played out in the 1980s and early 1990s with loan melas and debt-waivers are only two recent examples among innumerable others where huge damage has been inflicted on the national economy and most especially on the poor, through such misguided and corrupt giveaways. The notion that land is the only real security for the poor is an old fashioned myth that becomes ever more unreal with the passage of time. Over-emphasis on "equity" at the cost of "eficiency" has steadily brought down India's ranking on a global scale by whatever yardstick one might wish to apply. It has, moreover, brought social instability as a result of unfulfilled aspirations. The "equity" argument must be harnessed to reality.

The Rajasthan Government made no allotments for about 18 months while it rethought the policy issue. It then decided to go ahead with up to 50 per cent special allotments and, with better synchronisation of activity as between the engineering, colonisation and command area development wings, aim to achieve a target of 100,000 allotments per annum in due course. The project could be expedited if additional funds are made available and the completion date advanced by five or six years to 2000. For this to happen, the rate of settlement will need to be even further improved. Restrictions on resale of land will also need to be relaxed.

Indeed, some would even argue that if land were sold in open auctions, not merely would allotment and settlement be speeded up but the additional funds required for completion of the Project might more easily be mobilised. No speculator will spend a fortune in buying land in the IGN Stage II command and "uncommand" without being ready and able to develop it rapidly in order to earn a quick return on his investment. One has only to look at some of the fields and mandis in Stage I, such as Gharsana, to know how much prosperity has come to the area. There has been phenomenal capital appreciation both for land and commercial plots and people know they are sitting on a gold mine. Success did not come easily and certainly entailed hard effort and risk. But the opening of Rajasthan's "new frontier" has more than paid off.

The State should devise means of getting a share of these profits through appropriate cesses and taxes, including a betterment levy which was talked about in India in the early years and even legislated, but never collected. Past experience need not exclude a fresh attempt, better devised, to garner part of the unearned increment, especially if this could somehow be linked in part to reinvestment in local agro-processing and in the infrastructural and industrial development of the region. In Gharsana, land auctioned in 1969 for Rs 6000 per ha, is now selling for Rs 3 lakhs per ha. Shops in the mandi that were auctioned for Rs 24,000, now fetch Rs 10 lakhs. Business is brisk and some of the grain dealers measure their turnover in excess of a crore of rupees. In Stage I, the average price of developed land per murabba (6.32 ha) is around Rs 10 lakhs. At this rate the market value of the nearly 600,000 ha of previously desert land developed at a cost of Rs 1300 crores would be of the order of Rs 10,000 crores.

Pong Dam Oustees

Another issue widely agitated in relation to settlement, pertains to the plight of the so-called Pong Dam oustees. The very term "oustee" is peremptory, arrogant and derogatory and is an altogether unhappy term in the official vocabulary. Displaced person would be a more appropriate and accurate terminology.

Since the surplus waters of the Beas were largely allocated to Rajasthan, those displaced by the Pong Dam reservoir in Himachal were assigned for relocation and rehabilitation in the IGNP command, Stage I. The charge is that this commitment has not been honoured and that a substantial number of Pong Dam oustees have been denied resettlement or harassed and ejected in one way or another. While the Himachal government has taken up cudgels on behalf of the Pong Dam oustees, the Rajasthan Government feels that it has done the best it can and that the problem lies elsewhere. Given these sharply divergent views, it is necessary to take a closer look at the matter.

Work on the Pong Dam at Talwara in Punjab commenced in 1961. Just over 30,725 ha of land were acquired in the Haldoon Valley of Kangra district which was part of Punjab until the reorganisation of the state in 1967 when it was transferred to Himachal. Over 95 per cent of the acquired land falls within the reservoir which extends 42 km and displaced nearly 30,000 families. As a result of inter-State negotiations it was decided that those displaced should be given the option of rehabilitation in the IGNP command. The Rajasthan Colonisation (Allotment of Government Land to Pong Dam Oustees in the Rajasthan Canal Colony) Rules were promulgated in 1972. An area of 225,000 acres was specifically reserved for the Pong Dam allottees in the IGNP command.

Many of those displaced rehabilitated themselves locally by purchasing adjacent lands in higher reaches of the valley from out of the compensation they received. Of the rest, 16,100 families were identified as being eligible for resettlement in the IGNP command. Only those with up to 20 standard acres, of which 30 per cent or more had been acquired, were eligible provided that their residual holding did not exceed a ceiling of 30 standard acres and their names were listed in the land records as on March 31, 1961. Tenants, landless labourers and artisans were precluded from allotment of agricultural lands and were only entitled to house sites. For good measure, couples still within the reproductive age group and having more than three children born within the previous 10 years, were disqualified on family planning considerations.

The Himachal Government was required to appoint a certifying authority for the purpose of these rules before whom the oustee had to file an application with an identity card within four months of receiving compensation for the lands acquired from him. Once certified as eligible, the oustee had to submit an application with an affidavit to the allotting authority in Rajasthan. The Colonisation Commissioner, Government of Rajasthan, was authorised to relax the time limit but could write back to the certifying authority in Himachal for clarifications if necessary. Once the Colonisation Commissioner passed final orders, the oustee was required to deposit the requisite instalment in the designated treasury office and take possession of the land within 45 days of receiving an allotment notice.

On the completion of this forbidding drill, the Rajasthan Government approved 9195 applications and made the necessary allotments. The remaining 6905 eligible families either did not apply or could not pass the verification test. Allotments were made subject to the provisions of the Rajasthan Colonisation (General Colony) Conditions, 1955. Under these, the initial allotment was to be on a gair-khatedari basis with no rights accruing to the allottee for 20 years, and then only on payment of the full price of the land and all other dues. During this period, transfer or alienation of the land was barred, provided that if an allottee had paid the full price within 10 years he might be permitted to sell the property. The Rules further provided that within six months of the date of possession, the oustee must commence living permanently in the chak abadi to which his land was linked. Further, he must cultivate the land through personal, family or hired labour under his personal supervision. Widows, handicapped persons, students and members of the armed forces were exempted from this requirement. Finally, the allottee was required to bring the entire culturable area allotted to him "under the plough" within six months of the allotment and "fully utilise" the irrigation water made available. Breach of any of these terms would render the allottee liable to have his allotment cancelled.

How did things work out? Out of the 9195 allotments initially made, 6658 were cancelled on grounds of not taking possession of the murabbas within the stipulated 45 days, non-payment of instalments when due, not doing personal cultivation, sub-letting, or benami sale. In the result, only 2537 allotments had survived by 1980. The allottees pleaded procedural harassment, corruption, lack of facilities at the new sites, remoteness, encroachment, and threats from vested interests. They no doubt also faced an element of culture-shock, having moved from the cooler clime of lower Himachal to a hot, dusty and, at that time, inhospitable desert tract in Rajasthan. The Himachal Government too felt aggrieved and said that even without doubting the intent behind the Rajasthan rules, they had in effect worked as a deterrent to the settlement of the Pong Dam oustees.

Fresh negotiations between the two Governments resulted in a new Memorandum of Understaning in September 1981. The cancellations were to be reviewed and those who had failed to apply for resettlement in the initial round were permitted to file fresh applications if they were in hardship and without any alternative means of livelihood. It was agreed that those who had settled in the IGN command and had genuine problems with infrastructure and lack of amenities should be assisted, while those who had illegally sold or transferred their lands deserved no sympathy. For the rest, the Allotment Rules were amended to extend the period for taking possession from 45 to 90 days; cultivation by agents or managers was to be permitted; and cultivation was to commence within three years with 50 per cent of the command land irrigated the first year, the other 50 per cent the next year, and the uncommand land in the third year. Defaulted instalments were re-scheduled with six per cent interest. All encroachments were to be removed. Uncommanded areas were to be exchanged for commanded areas. Residence within murabbas instead of only in the chak abadis was allowed. Pending and time-barred cases were to be reviewed under the new Rules and applications were invited for restoration of resumed lands.

The outcome of the so-called Phase II Review and Allocation by October 1991 was that 4734 applications were received and of these as many as 4510 allotments were restored and possession was actually delivered in 3277 cases. Only 224 applications were rejected and 1233 were pending. Thus, by the end of 1991 only 5814 out of 16,1000 eligible Pong Dam oustees had actually resettled in the IGN command.

Rehabilitation is at the best of times a complex process. It is greatly complicated when inter-state resettlement is involved and even more so when the agro-climatic and socio-cultural differentials are as great as those prevailing in the case of Pong Dam oustees being sought to be resettled in the Rajasthan desert. The initial concept of land settlement was faulty and should have been a residual rather than a prime component of resettlement and rehabilitation. The notion of giving land for land as the sole or best means of rehabilitation is questionable. There was, of course, no dearth of land on offer in the IGN command. This land was being offered as Rajasthan is the beneficiary state and there was no land available in Himachal. Training for employment in other occupations was scarcely explored though this is increasingly becoming the most viable option for rehabilitation.

However, given a policy decision to rehabilitate the Pong Dam oustees in Rajasthan, more could have been done to acquaint them with the implications of this option. The Rules were needlessly stringent and perhaps mechanically applied by officials of another State who might reasonably be supposed to be less sympathetic to the claims of 'outsiders'. The alternative of locating a team of Himachal officials and NGOs at the resettlement sites so as to smoothen the transition and sort out bureaucratic hassles does not appear to have been pursued. The Rajasthan Government constructed over 4000 dwelling units between 1973–76 for the oustees, few of which were apparently ever occupied by those for whom they were intended. Squatters moved in later. Many oustees, finding the proposition more difficult than they had bargained for, and being totally unfamiliar with the kind of reclamation operations required to level and break the land, took the easy way out of selling their allotment under the table and returning home to Himachal. They sold the land at distress rates and later, when the area had developed and land prices started soaring, felt cheated and clamoured to return. Meanwhile communications had improved and Kangra is now less remote from

Stage I or II of the IGNP than before.

The issue has got politicised and at one time a section of embittered oustees demanded stoppage of Pong power to Rajasthan. This is a humanitarian question and can still be reasonably resolved with a more innovative approach by both governments.

Shelterbelts and Plantations

Trees are a blessing anywhere. Never more so in the scorching desert with raging sandstorms and mobile dunes that bury everything in their path. The sand threatens the canals, roads and fertile lands. Shelterbelts, road and canal-side and abadi plantations and sand dune stabilisation measures are therefore essential. Work on a variety of afforestation, pasture development and fuelwood plantation programmes was accordingly taken up quite early on in the IGN command and has steadily gathered momentum. Until March 1992 some 146,000 ha of afforestation under these various heads had been completed. About 37,000 ha of this is in Stage II which is ultimately planned to have 362,000 ha under plantations.

The canal network runs to several thousand km and in Stage I strips of land on either side were taken up for plantation with a width of 100 m along the Main Canal, 50 m along branch canals, 30 m along distributaries, and 15 m along minors and subminors. A project for planting a two-kilometre belt along the left bank of the Main Canal in Stage II is part of a 50,000 ha forest and pasture development programme with support from the Eco-Development Task Force of the Territorial Army. About 800,000 saplings are being planted each year with irrigation and handed over to the forest authorities after three years. The lush foliage is there for all to see. Sand dune stabilisation programmes are also visible with tawny banks literally bio-crated in bands of green and deep grey wire-mesh. OECF Japan is assisting these programmes. New abadis have 12.5 ha of command land earmarked for fuelwood plantations on which a variety of fodder and minor timber species have been grown.

In waterlogged reaches and areas of seepage along the Main Canal, bio-drainage plantations have been raised, some of them salt-tolerant species wherever there are shallow brackish aquifers. Species such as eucalyptus, khejari, Prosopia juliflora and certain accacia varieties thrive.

Pasture development is being taken up on both command and 'uncommand' land with limited irrigation by camel carts to ensure that the plants are well established. Fencing of degraded areas has also resulted in regeneration.

By 1991, over Rs 213 crores had been spent on the afforestation package. But on completion of the IGNP, the afforestation programme in expected to yield about Rs 126 crores of forest produce annually. Seedlings are being distributed to farmers for planting along field borders. These too will protect farmlands from inundation by sand and make their own contribution to the economy and ecology of the region.

The extension of the canal system and plantations in the desert is creating new ecological-niches that have begun to attract a variety of birds, including duck, cranes and other aquatic species, small game and black buck. The desert is coming alive.

Animal Husbandry

The Western desert, though inimical to anything but the hardiest crops, has always sustained nomadic pastoral communities which have practised animal husbandry. Rathi and Tharparker are famous desert cattle breeds like the magra sheep. Bikaner has been an important dairy and ghee production centre. The critic's lament is that, despite the centrality of animal husbandry to the economy of Western Rajasthan, the pastures, common grazing lands and gocher-orans (sacred pastures) have been allowed to be encroached upon, distributed and diverted, robbing the community of traditional fodder banks. These commons are said to have "protected the environment and established a balance between man and animal and man and circumstances . . . and involved a process of community decision making". These have virtually gone. (URMUL Trust, 1992) The community has lost control over the commons while the State has been indifferent to their fate.

The fact is that despite sparse numbers, both population and livestock have been increasing and overgrazing has aggravated desertification and degradation of the environment. Unthinking populism in the name of "equity" has been partly responsible for the loss of the commons. The IG Nahar is now changing the farming pattern from pastoralism to settled, irrigated agriculture. However, in view of many considerations, including soil and irrigability factors, animal husbandry, and pasture and dairy development are being encouraged. Crop farming is displacing sewan grass but fodder cultivation is being encouraged. The Rajasthan Cooperative Dairy Federation has 16 milk unions with 10 dairy plants. The IGN command is covered by the Jodhpur and Bikaner dairies. The Bikaner dairy plant was established in 1978 with a 100,000 litre per day capacity, subsequently increased to 150,000 lpd and now proposed to be doubled again. The current RCDF capacity statewide is 1.2 million lpd. Average annual rainfall in the Bikaner milkshed is around 28 cm but the rains are erratic and was persistently poor for several preceding years. In bad times herdsmen and their cattle migrate. Less water entails less fodder of a lower nutrient value and hence lower milk yields and less calving. The Bikaner milk union had 430 societies some years ago. But only 270 were active in 1993 as entire villages had migrated during the preceding drought years and have not yet returned.

A Rathi cow yields on an average eight lpd with the best yields rising to 16 lpd per animal. But in a herd of 100, only five or six animals may attain that output. With tractorised ploughing, sewan grass is disappearing but farmers with irrigation keep at least one bigha under fodder. Fewer animals are being reared but these are of better quality. As a result, milk production, though less than before overall, is rising in the irrigated tracts to about 1000 litres per year of 250 milk days with an 18 month calving cycle. Farmers tend to retain about 30 per cent of their output for household consumption as milk is their main protein and fat source.

A Rathi bull breeding project has started. Within the irrigated areas, stall feeding is coming into vogue. But outside the command, cross-breeding has failed as the herds are let out to graze and cows cannot be serviced in time when in heat. Artificial insemination is gaining popularity in the command. The Canal is changing the nature of animal husbandry. Properly planned, it will certainly be for the better.

The advent of irrigation has introduced the possibility of developing fisheries in the IGN command. The diversion of flood waters into the Ghaggar depressions has led to a thriving contract fishery. Fingerlings are released in the depressions, the catch varying from year to year in relation to the quantum and duration of ponding. 1991-92 was a record year but in other years production has been down to a few hundred tonnes. Pisciculture can be developed on a more extensive scale in excavated canal-side fish ponds and even in cage fisheries in the larger canals. Brackish water and drainage flows can also be led into ranns or other depressions and used for raising brackish water fish. There is much scope for innovation.

Quality of Life

Introducing water into hitherto dry areas does carry the risk of waterborne diseases such as malaria. These can be contained through a variety of preventive measures and by health education. On the other hand, water scarcity is responsible for various ailments and the mere provision of adequate water can have a significant impact on general health and hygiene. Overall, it is reasonable to expect that the long term health impact of the IGNP will be very positive.

Rising agricultural production, incomes and employment favour trends towards better diets and an improved nutritional status, though evidence during the earlier transitional years could be mixed.

The greatest boon of the Project is, of course, the provision of drinking water. There are substitutes for many things but least of all water. The nomadic culture of the desert with periodic to frequent migrations of whole communities and livestock in "drought" years is self-evident. What this does to education, health and living standards generally is not always well understood. It constitutes a major hardship and handicap. Women and the girl-child are primary victims, for they spend the best years of their lives, day after day, and several hours a day, just walking wearily to fetch water, fodder and fuel in arid surroundings. The cost of this is never computed and must be enormous. Many social customs probably derive from a poverty-inducing environment. One of these is the Western Rajasthan celebration of Akha Teej or Akshay Tritya during which mass infant marriages or betrothals are solemnised. This, according to some sources, reduces the cost of weddings and of dowry and engenders a sense of security among nomadic groups whose paths cross infrequently.

Until the IG Nahar came, traditional water sources and water harvesting devices were relied upon until bad years compelled migration. The Lathi and Dabla aquifers constitute an oasis. But this is fossil water with little recharge and will be exhausted in due course. The other water sources draw on perched aquifers.

The IGNP has reserved 0.93 MAF of Rajasthan's allotment of Ravi-Beas waters for drinking water supplies (0.745 MAF), industry and power generation (0.13 MAF), and afforestation (0.050 MAF). The drinking water component is going to be Western Rajasthan's life support system, providing ultimately for a population of nearly 17 million by 2031, including supplies for the principal cities in the desert and a sprawling rural population, all of whom have been rapidly running out of water. The six lifts that form part of the Project will convey supplies from the canal network over leads of 60 to 200 km with lifts of 60 to 239 m to service this population. The conveyance system will consist partly of open channels and partly of pipelines, the latter being five times more energy-intensive than the former owing to friction losses. Yet such is the threat of choking of channels by sandstorms and the depth of cuttings required in some segments that piping the water is considered the better alternative. Canal-side plantations and protective shelterbelts are also planned. Deliveries have commenced and will be augmented steadily as the system is developed. A series of intermediate pumping stations have been or are being constructed to lift the water in stages. The power supply, now sometimes erratic, will require to be stabilised. Lignite and gas deposits in the region are sufficient to sustain 420 MW-size power plants based on gas at Kapoordi in Barmer district and Ghotaroo in Jaisalmer district, and on lignite at Barsinghsar and coal at Suratgarh in Bikaner district. The requisite cooling water from the Canal is being reserved for these thermal units as well as for the agro-processing and other industries they are likely to attract. Some 48 MW will ultimately also be generated in micro-hydel stations on some canal falls. Even small falls should be sought to be exploited to operate local processing units and cold storages.

The water supply systems are being developed by the Rajasthan Public Health Engineering Department. In Jodhpur, a city of 650,000, inhabitants consume 130 lpd of water drawn from a number of groundwater sources within a radius of 60 km. The criticism is that with the tapping of these sources, the traditional water harvesting system that Jodhpur once relied upon was abandoned. The city had moved to this present site 600 years ago from Moderna, the former capital of Marwar, which had run short of water. Now Moderna's fate is felt to be overtaking Jodhpur. Groundwater has been recklessly mined and, with the city's expansion and a seemingly plentiful supply of water, farmers have in recent years unwisely taken to cultivating water-intensive crops. More distant villages, whose groundwater supplies have been tapped to serve the city, are up in arms at this "robbery". (Shankar, Uday. July 1993)

The Jodhpur Lift, 179 km of canal and 25 km of pipeline, was sanctioned in 1984 and will pump 289 cusecs to service an ultimate population of 1.8 million with 250 million lpd. Barmer, Jaisalmer, Bikaner, Kolayat, Pokharan, Nagaur, Phalodi and Lunkaransar are other cities that will draw water from the IG Canal. This is an enormous undertaking and would be justification for the IGNP in itself. Nevertheless, the caution about not discarding and even reviving traditional water harvesting and storage methods, avoiding unwise cropping patterns and rash groundwater exploitation, is well taken. At the same time it would be an exaggeration to suppose that such traditional methods by themselves could ever have sufficed to provide the necessary water supplies for the growing number of people and livestock in the region. And, if as a result of the greening of the desert, sand dune stabilisation and dust suppression, the changing micro-climate over the next 30–50 years coaxes a little more rain over this region, which already has a high atmospheric humidity, this could be an additional gain of lasting benefit.

Signs of Transformation

What was once a distant vision is transforming the desert. There is a long way to go before the Project is completed. There is need to correct some imbalances such as waterlogging in certain patches and to improved drainage, the rate of settlement and coordination between different Departments. But the outcome is no longer in doubt and could be hastened with the provision of additional funds, which should be a national priority. In the transitional period there are and will be problems. The residual Pong Dam oustees should perhaps be rehabilitated through an offer of self-employment opportunities rather than on the land. Those moving from nomadism to settled agriculture and a new pattern of animal husbandry will need help and guidance through social extension services which NGOs would be well placed to provide. Ills such as alcoholism and opium smoking need to be tackled early. Otherwise the signs of change in the quality of life are evident: access to water, roads, electricity, better housing, new dresses.

Incomes are rising and employment is expanding and getting more diversified. The Stage I project completion report indicated that net average household income in the project had gone up from Rs 7671 per annum in the neighbouring unirrigated tracts to Rs 16,792 in the command. This will rise further. Agriculture contributed 47 per cent to total household income compared to 18 per cent in the unirrigated areas. Indebtedness in the command had risen by seven per cent as against 51 per cent in the non-commanded areas, while household employment had increased from 589 to 640 mandays per annum. (Ramanathan, S. and Rathore, D.S. 1991) Education and health indices were disappointing and constitute two areas deserving of far more attention.

Water rates should be priced up and drinking water, now being given free, should be charged, even if initially only at a nominal rate. Free economic services are no part either of equity or welfare and are sooner or later going to be unsustainable. Betterment, wealth tax and an agriculture income tax in some form appear necessary and early introduction would preclude non-payment hardening into an imagined right as so often happens.

The IGNP will pay its way economically and has already done so in any overall national social accounting, including its value as a formidable defence ditch along a long stretch of open border. Above all, it has softened the rigours of the desert and is making it a happier home for millions. Chinkaras gambol and peacock dance beside the Canal.

REFERENCES

1. Mishra, V.N, Evolution of the Pattern of Human Settlement in the Arid and Semi-Arid Regions. *Desertification and Its Control,* ICAR, New Delhi, 1977.

2. Savithri, R., Studies in Indian Archaeo-botany with Remarks on the Socio-Economic Environment of Indian Proto-Historic Cultures. PhD thesis, Lucknow University. Cited by Vishnu- Mitre in *Desertification and Its Control,* ICAR, New Delhi, 1977.

3. Krishnan, A., A Climatic Analysis of the Arid Zone of North-Western India. *Desertification and Its Control,* ICAR, New Delhi, 1977.

4. Tod, James, *Annals and Antiquities of Rajasthan,* Volume II. Routledge & Kegan Paul. London, 1832.

5. Anon., *Development of the Aravali Hill Region.* Paper presented at the Aravali 2001 A.D Conference organised by the Ubeshwar Vikas Mandal, Udaipur. Jan 31, 1986 to Jan 2, 1987.

6. Ramanathan, S. and Rathore, M.S., *Equity and Productivity Issues in IGNP - A Review of the Literature.* Institute of Development Studies, Jaipur. December 1991.

7. WAPCOS, IGNP-Stage II Draft. Final Feasibility Report. New Delhi, 1993.

8. Indira Gandhi Nahar Board. *Profile of Indira Gandhi Nahar Project.* Government of Rajasthan, Jaipur. 1993.

9. Commissioner, Area Development IGN, Monitoring of Water Table : Drainage Trials in Command of IGNP Stage I. Technical Report-IV. Bikaner, 1991.

10. URMUL Trust, *The Nahar Yatra: A Report on the Indira Gandhi Canal.* Bikaner, August 1992.

URMUL Trust, *Commons and Canal: A Comment.* Bikaner, 1992.

11. Shankar. Uday, Groundwater Disappearing Act. *Down to Earth,* New Delhi, July 15, 1993.

4

The Tehri Controversy

Origins

THE Ganges Canal, commissioned in 1862, was a monumental engineering feat for its time. Though the Western Jamuna Canal dates back to the 14th Century, there was really no Himalayan storage in the entire Gangetic basin until the Ramganga dam was completed in 1970. The Ganga itself was initially annually diverted through temporary bandhs at Hardwar for some decades until the Bhimgoda barrage was finally constructed. Even then, a small gap was retained in the structure to ensure that the free flow of the holy river was not totally obstructed.

The Ganga irrigation network has since expanded, with further barrages at Bijnor and Narora, to feed the Madhya and Lower Ganga Canals. These provide protective irrigation from natural flows which are insufficient for the more intensive hybrid seed-fertiliser technology to which farmers have adapted. Emerging water shortages in the Ganga Canal, now being modernised and augmented, and increasing shortfalls in peaking power soon turned attention to harnessing the monsoon flows of the Ganga catchment above Rishikesh to meet both requirements.

The Ganga takes that name at Devprayag, 40 km above Rishikesh where the Alaknanda, rising near Badrinath, joins the Bhagirathi which has its origin at Gaumukh, the snout of the Gangotri glacier. The Alaknanda carries two-thirds of the Ganga's flow at Devprayag, and the Bhagirathi the remaining third. But the descent of the celestial Ganga is associated with the sage, Bhagirath, who broke the fall of the mighty torrent in his matted locks and brought it gently down to earth.

A dam on the Bhagirathi was conceived in 1949. It was, however, not actively pursued until 1961 when four sites were investigated and Tehri selected. Further investigations were conducted and technical formulations developed in consultation with a number of Indian and foreign experts. In view of high seismicity and a suspected river bed fault right under the dam site, a rockfill dam was preferred in place of the concrete structure originally contemplated.

A detailed project report was submitted by the U.P. Irrigation Department in 1969. This envisaged a dam more than four times wide (1125 m) as it was tall (260.5 m above its deepest foundation), a massive wedge fanning out to plug a narrow canyon. The benefits included irrigation and 600 MW of installed generation capacity. Taking note of the rising load curve, a revised project report in 1979 envisaged raising installed capacity to 1000 MW. Another 1000 MW of peaking power was to be generated through reversible turbines at a second power house with the creation of a balancing reservoir following the construction of a 103.5 m concrete dam at Koteshwar, 23 km downstream. Pumped storage entails the utilisation of an available surplus of non-peak base-load thermal energy to return water earlier passed through the turbines back to the main storage for subsequent generation of power when required to meet peak loads. With an installed capacity of 400 MW at Koteshwar, the total potential of the Tehri Dam complex was augmented to 2400 MW.

The Tehri Dam complex is further intended to bring an additional 270,000 ha under irrigation and stabilise existing irrigation over 600,000 ha in the Ganga-Yamuna doab.

The expected benefits in terms of 5071 m units of energy, irrigation, pisciculture, tourism, 300 cusecs of drinking water supply to Delhi, employment and income-generation were considerable. The total cost was estimated at Rs. 2036.6 crores and the project was to be fully commissioned by 1994-95. Though no provision was made for flood control as such, some cushion will be there incidental to the impounding of monsoon flows in the Tehri reservoir with a gross storage of 3540 m cu m or 2.87 MAF.

The Tehri Project will regulate a catchment of 7511 square km containing the headwaters of the Bhagirathi and Bhilangana rivers which meet near Tehri town, itself but a kilometre above the dam site. The reservoir will stretch back 44 km up the Bhagirathi valley towards Uttarkashi and 25 km up the Bhilangana valley, submerging an area of 4200 ha including Tehri town. Of the submergence area, 1600 ha is being farmed and an equivalent area is forest land. The remaining area is uncultivated. Together with the areas affected by the Koteshwar dam, the project colony, approach roads and service areas and the land required for New Tehri Town, 43 villages will be fully submerged or acquired, another 69 partially submerged and a further 10 villages otherwise affected, altogether displacing some 13,500 families. While much criticism of the project is common to several other large dam schemes, the Tehri controversy has uniquely revolved around fears that the dam will break, given the high seismicity of the site, releasing a devastating flood liable to sweep and destroy everything in its path. The safety of the dam was always questioned but the intensity of feeling about the alleged seismic dangers reached a crescendo with the Uttarkashi earthquake in October 1991 which took a grievous toll of lives. More on that later.

The Planning Commission approved the Project as originally conceived in 1972 and the Uttar Pradesh Government accorded its administrative approval four years later. The Chipko movement had begun to spread environmental awareness in Garhwal and concerns about dam safety, the life of the dam on account of presumed heavy siltation, the loss of some of the best flatlands in an otherwise mountainous landscape, and rehabilitation issues came to be combined in the opposition to the project. Meetings of local gaon sabhas (village councils) and the zila parishad (district council) were organised in 1977 leading to the establishment of the Tehri Bandh Virodhi Sangharsh Samiti (or Anti-Tehri Dam Struggle Association) the following year. Protests, demonstrations and fasts were staged to block excavation of the diversion tunnels and a petition was presented to Parliament. The Lok Sabha was dissolved in 1979 before the Committee on Petitions could conclude its inquiry. The submission of the revised project report incorporating the Stage-II features revived the agitation.

In a counter move the Central Government announced the constitution of a Working Group to assess the environmental impact of the Tehri Dam under S.K. Roy (1980).

Working Group

The Working Group was divided from the start with differences sharpening as time went on. An interim report was submitted in May 1980 in which the environmentalist members pleaded very inadequate data on a large number of parameters which they felt must be made good by the authorities concerned before they could frame their recommendations. Issues of seismicity, reservoir rim stability, siltation rates, catchment area treatment, rehabilitation and related matters were raised. The engineer members felt that most of these questions had been adequately investigated and that undue notice was being taken of vague assertions and misperceptions.

A large number of further studies and additional data sought by the environmentalist members was not forthcoming to their satisfaction. In the circumstances, the Chairman rejected the draft report prepared for the group and unilaterally submitted a report on his own in August 1986 with a covering letter which other members read as a sharp dissent from their views. While the need to augment national energy potential by harnessing the bountiful Himalayan water resources was recognised, the Chairman's concern centred on "the known instability of the hillsides, the extensive deforestation and the location of the TD in a highly sensitive seismic zone". On this last issue in particular the Working Group was divided on the emphasis placed on the "technological drive for development irrespective of critically important human and environmental considerations . . . The environmentally-oriented members question the assumed technological infallibility. They consider that the inadequate data presented only adds to the concern that there is a major risk element . . .". The Chairman expressed his personal conviction that the Tehri site was environmentally unsuited for a high dam and that work on the project be stopped pending a re-examination of design parameters to see how the expenditure already incurred "can best be incorporated in a revised project either with a low dam or through a modified run-of-the-river project". Shortly thereafter other members of the Working Group wrote to the Government disclaiming the Report submitted by the Chairman.

The breach was complete. The Ministry of Environment and Forests on the one hand and the Department of Power along with the Central Water Commission on the other appeared equally divided on the Report. To allay public doubts, the CWC appointed a committee of scientists and engineers to evaluate the

Dam on considerations of seismicity and safety.

Setting aside the seismic aspect for the moment, the Working Group had made a number of other observations and recommendations. One of its major concerns was with regard to siltation and catchment area treatment. In its interim report it had wished road building to be limited and mining or quarrying to be prohibited within the catchment. It felt it imperative to assess the environmental impact of the Tehri Dam not in isolation but in conjunction with several other schemes whether completed, ongoing or planned to harness the waters of the Ganga above Rishikesh which should be treated as a single planning unit. To this end it recommended the establishment of a Ganga Management Authority. Other matters raised related to displacement and rehabilitation, health impacts, command area development, pisciculture, flora and fauna, tourism, ferry services across the reservoir, and the need to keep people informed through public relations. It called for a sub-basin approach to water administration, continuous monitoring of water quality, greater instrumentation of the region to measure seismic patterns and impacts, improvement of soil and water conservation on agricultural lands within the catchment, including lining of channels, improved terracing, development of horticulture and dairying, watershed management, social forestry and landslide stabilisation measures. Command area development was not to be neglected. It advocated the creation of a conservation corps to involve local communities and the large number of ex-servicemen belonging to the region in these tasks. It examined the need for providing a specific flood cushion but found on expert advice that this would be of little use in moderating 25 and 50-year return floods at Kanpur. (Working Group, August 1986) The Chairman doubted the financial viability of the Project.

Alternative Schemes

Opposition to the Dam had from the start been linked with alternative proposals. In 1981 it was proposed that the upstream Maneri Bhali hydel project's tail-race waters could be diverted through a contour channel and the head difference be utilised at Tehri to generate power. Similarly, the tail-race waters of a low dam at Tehri could be carried in an open channel to Devprayag and the drop utilised for power generation. It was admitted that the power so generated would be less than that proposed under the Tehri High Dam project but it would be steady power which might be augmented by similar systems in the Garhwal Himalaya, as for example on the Pindar River. (Naqvi, March 1981)

In 1985 it was reported that the Sangharsh Samiti had suggested that water from a low dam at Tehri be diverted through an open channel on the right bank of the Bhagirathi to Bhason, 8 km downstream of Tehri, where a power station could be located. Thereafter, another open channel could take the water at Pindaras, near Devprayag, for further generation of power. "Such a scheme will preclude the need for inundating vast areas and will also bring the fruits of development to the people of Tehri-Garhwal." (Lal Bahadur Shastri Academy, October 1985) A section of the Working Group, headed by the Chairman, also favoured a low dam or a modified run-of-the-river scheme.

Alternatives to the Tehri High Dam were, indeed, considered by the Central Water Commission and Central Electricity Authority in 1982 at the instance of the Ministry of Irrigation and again around 1988. Three alternative storage schemes were examined. The first of these envisaged a 35 m lower dam at Tehri with only half the storage proposed under the High Dam scheme. Tehri town would still be submerged but five villages would be spared full submergence and 10 villages partial submergence. The power output for Stage I would be 2354 m units as against 2910 m units planned. The unit cost of generation would be higher, the irrigation potential half and the cost-benefit ratio distinctly lower. The second alternative looked at was a dam at Dobata on the Bhagirathi, above its confluence with the Bhilangana, 30 m lower than the Tehri Dam and saving Tehri town. The storage would be much less as the Bhilangana would be excluded. The irrigation and power benefits and the cost of generation all compared unfavourably with Alternative I. A third alternative envisaged two separate dams on the Bhagirathi and Bhilangana three km above their confluence, with a tunnel linking the two reservoirs. Tehri town and three villages would be saved. The economics and benefits would be an improvement on Alternative II but would not match Alternative I. None of these schemes therefore would be as worthwhile as the proposed Tehri

High Dam.

Three possible run-of-the-river schemes were also examined that would utilise the head available in the reaches of the Bhagirathi and Bhilangana rivers submerged by the Tehri High Dam. These schemes were the Dharasu-Chham hydel scheme.(90 MW installed producing 481.8 m units per annum at 85.5 P per unit); the Chham-Tehri hydel scheme (120 MW installed, producing 632.4 m units per annum at 81.5 P per unit); and the Bhilangana hydel scheme (99 MW installed producing 419.6 m units per annum at 95 P per unit). The combined cost of all three schemes worked out to Rs. 1208 crores (at 1987 prices) and they would generate 1552.2 m units at 87 P per unit. This compared with a cost of Rs. 1455.73 crores (1987 prices) for Tehri High Dam Stage I, inclusive of irrigation and catchment area treatment benefits, and generation of 2910 m units at 43 P per kWh. The run-of-the-river schemes would forgo irrigation and incidental flood control benefits, peaking capability and water supply to Delhi. Studies were earlier undertaken to vary the height of the dam by 15 m above and below the projected level of the Tehri High Dam which is to stand 260 m above its deepest foundations. Every study indicated that the final project optimised the sum total of all parameters and it is this proposal that is being implemented. (Asthana and Jain, 1991)

Court Ruling

Impatient over the delay in the submission of the final report of the Working Group, which they were closely monitoring, the Tehri Bandh Virodhi Sangharsh Samiti filed a writ petition before the Supreme Court in 1985 against the State of U.P. and others. It challenged many aspects of the Tehri High Dam and particularly its safety.

The Petition said the dam site was risk-laden, the rock formation weak and the region too critically stressed and seismically active, with numerous active thrusts, faults, and shear zones, engendering deep-seated fears about the consequences of a devastating earthquake. "The chances of failure of the dam are too great and real and, if that evil occurs, all habitations like Muni-ki-Reti, Rishikesh, Hardwar, et cetera, downstream of the dam will be wiped out of existence, causing a disaster unprecedented in the history of man. It is claimed that the Respondents have no right in law or otherwise to make such massive inroads into the fragile eco-system of the Garhwal Himalaya and create hazards to the life and property of the people or to sink valuable lands of the nation under water and deprive posterity of the benefits thereof for some assumed temporary benefits. It is claimed that even if the dam does not fail in the meantime, the useful life of its reservoir, in view of the high sediment load carried by the rivers, can by no stretch of imagination be more than 15-20 years." The Project could not be considered a "reasonable engineering risk and is wholly incommensurate with the community benefits expected to accrue from it". It advocated taking up alternative run-of-the-river projects.

In sum, "the Tehri Dam Project is a very dangerous venture, it is technically infeasible, geologically a blunder, economically unsound and environmentally disastrous. It is not a project of development but a scheme for destruction . . . a criminal act and an unpardonable ecological sin". The Court was asked to forbid the Project. There could not have been a stronger indictment. (TBVSS, 1985)

The Director, Environmental Cell of the Indian National Trust for Art and Cultural Heritage (INTACH), filed an application as intervenor in support of the Petitioners and submitted written submissions "on factual, scientific and legal grounds". These held the chances of a devastating earthquake arising in close proximity to the dam as very great in view of "the river-bed fault which crosses the dam site". It asserted that the Tehri Dam jeopardised the Right to Life under Article 21 of the Constitution and described the project as anti-people, an assault on the well-being, culture and eco-system of the hill people, and an irrecoverable financial loss. The indisputable fact of the matter, it said, is that the real beneficiaries of large scale dams and water development schemes "have invariably been large multinational companies, the urban elite of the Third World and the politicians" who commission these projects. (Jayal, 1986)

The Central and U.P. Governments countered these arguments in elaborate detail citing the various studies made. The Court passed its order in Novermber 1990. It observed: "The questions relating to the design of the dam, the seismic potential of the site where the dam is proposed to be constructed, and the various steps which have been taken for ensuring the safety of the dam are highly intricate questions relating to science and engineering. This Court does not possess the requisite expertise to render any final opinion on the rival contentions of the experts. In our opinion, the Court can only investigate and adjudicate the question as to whether the Government was conscious of the inherent danger as pointed out by the Petitioners and applied its mind to the safety of the dam. We have already given facts in detail which shows that the Government has considered the question on several occasions in the light of the opinions expressed by the experts. The Government was satisfied with the report by the experts and only thereafter has clearance been given to the Project." The Court appreciated the Petitioners' concern for the safety of the dam but concluded that "in view of the material on record, we do not find any good reason to issue a direction restraining the Respondents from proceeding ahead with the implementation of the Project". The Petition was dismissed. (Supreme Court, November 1990)

The controversy, however, continued unabated and the tragic Uttarkashi earthquake in October 1991 gave it further impetus. Though various experts had considered the seismic factor once again in view of public fears, INTACH through its Director, Natural Heritage, filed another petition before the Supreme Court along with Shekhar Singh, an environmentalist, who had earlier been a member of the Environmental Appraisal Committee on River Valley Projects of the Ministry of Environment and Forests, which had also examined the Tehri Dam. The Petitioners cited the Uttarkashi earthquake, reportedly of a magnitude of 6.1 on the Richter scale, as having devastated areas proximate to the Tehri dam site. If this was so, then anxiety must mount in regard to the destruction likely to be caused by a far bigger M 8.5 earthquake that might shake Tehri dam. Secrecy was being maintained in respect of documents and the reasoning underlying decisions and therefore both the right to life and the right to know were involved. The petitioners prayed the Court to summon all records and documents, stay the construction of the dam and protect the people and ecology of the region. (Jayal and Singh, December 1991) The matter is pending and the Government is expected to challenge it on the ground that there has been no material change in the situation since the Court gave its earlier ruling in November 1990.

Environmental Appraisal and After

Having disposed of the litigation, one must return to the sequence of events. The Environmental Appraisal Committee headed by Dr. D.R. Bhumbla, a distinguished agricultural scientist, was asked to consider the ecological and social impacts, costs and benefits of the Project. It, however, went on to review the geological and seismic factors and the hazards these posed. Having done so, it unanimously concluded that the Tehri dam project did not merit environmental clearance and should be dropped (1989).

The EAC Report was evaluated by the Committe of Secretaries in March 1990. The view taken there was that the EAC should have confined its appraisal to subjects within its competence, leaving the safety aspects to be considered by those knowledgeable in matters of earthquake engineering and design and to the specialised and scientific organisations concerned with these disciplines. It nevertheless decided to turn over safety aspects to a High Level Committee of Experts. Pursuant to this, the HLC was constituted under D.P. Dhoundial, Director-General, Geological Survey of India, and five others. The Committee submitted its report in April 1990 stating the dam was safe. However, in view of some disagreement being communicated by one of the members, Dr. V.K. Gaur, an eminent geophysicist, the Government referred the matter back to the HLC. This Committee reiterated its confidence in the dam. Dr. Gaur formally dissented and suggested the matter be referred to an independent seismological expert of international repute.

The Government accordingly requested the Department of Mines to make such a reference, which it did to Prof. Jai Krishna, former Vice-Chancellor, Roorkee University, former President of the International Association of Earthquake Engineering, and former President of the Indian National Academy of Engineering. Dr. Jai Krishna endorsed the HLC report and found the Tehri Dam safe against the strongest expected earthquake in the region. Dr. Jai Krishna's impartiality was, however, impugned by INTACH which pointed to his earlier views and association with the project as factors indicating bias. But the Government felt that there was no cause to set aside his considered findings.

The Uttarkashi earthquake, however, provoked strong local reactions with Sunderlal Bahuguna, the Chipko leader who has long opposed the Dam, undertaking an indefinite fast at Tehri. This attracted wide media and political attention. The Prime Minister ordered a halt to blasting, as it was complained that the vibrations so caused could bring down buildings in Tehri already weakened by the earthquake, and promised a further review of project safety. This review was conducted by a committee of earthquake engineers, seismologists, geophysicists and other scientists representing all points of view. After much deliberation, it concluded in 1992 that the dam as designed is safe.

The Soviet Connection

In the midst of this imbroglio, another more mundane problem arose — finance. The Government looked for a partner to execute the project and found one in the Soviet Union. An Indo-Soviet Agreement was signed towards the end of 1986 under which the Soviets offered technical and financial assistance for the project, undertaking to execute the entire power component in both Stages as well as the Koteshwar dam and the transmission system on a turnkey basis. Apart from its offer of 1000 million roubles, the Soviet Union's participation was welcomed in view of its considerable experience in high dam construction in seismic-prone regions of Central Asia, including the 335 m Rogun dam (under construction) and the 300 m high Nurek dam (completed) in Tadjikistan. Soviet association and reappraisal of the Tehri dam design, it was felt, would allay misgivings about dam safety.

The Soviets laid out a whole series of parameters for study and testing in keeping with their standards and design practice and confirmed the feasibility "of a rock-fill dam and spillway at the site selected by the Indian side, as well as the principal layout and design solutions for the said structures". They too pronounced the dam safe after detailed study and two-dimensional dynamic testing. Construction was beginning to gain momentum when the disintegration of the Soviet Union and the Uttarkashi earthquake applied financial and physical brakes to the project. As successors to the contract, the Russians were anxious to carry out their commitments, but the political and financial disruption caused by the falling apart of the Soviet Union were to prove insurmountable in relation to Tehri. An exploratory drift having been driven below the river bed at the Tehri dam site, the Soviets confirmed that this ruled out "any possibility of the existence of a fault or major shear zone at this site". (Indo-Soviet Expert Negotiations, December 1987) Low water permeability or absence of seepage in the dam foundations was also confirmed, laying to rest yet another earlier surmise that had been a cause for concern. The dam crest level at 839.5 m and the average value of the slope of the Main dam at 1 : 2.5 m upstream and 1 : 2 m downstream were endorsed. That is, contrary to repeated assertions by critics despite official clarifications and contradictions, all major dam specifications remained unaltered. It was not recommended that the width of the dam at river bed level be extended from 1125 m to 1500 m, as attributed to the Soviet Chief Engineer (for Tehri), Alexander Fink. (Paranjpye, 1988) The Indian Chief Engineer, Tehri Dam, M.M.L. Khanna stated that Fink was being quoted "absolutely out of context" in this regard. (Khanna, January, 1988) A subsequent report captioned "USSR not keen on Tehri Project" (Rai, November 1988), evoked a rejoinder by Fink expressing "dismay" and describing as "baseless" the assertions made. (Fink, November 1988) INTACH, along with 54 other bodies around the world wrote to the Russian President, Boris Yeltsin, on the eve of his visit to India in 1992, not to revive funding for the Tehri project. Sunderlal Bahuguna added his mite by stating that Soviet experts were doubtful whether the Tehri Dam would store sufficient water and that irrigation would therefore only be possible once in three years. (Bahuguna, June 1992) Another myth. The campaign against the dam goes on. (*Bhagirathi ki Pukar*, February, 1993)

Environmental Clearance

Although the Ministry of Environment and Forests had initially gone along with the report of its Environmental Appraisal Committe in 1990, the over-ruling by the Committee of Secretaries of the EAC's observations on dam safety, caused it to relent on environmental clearance on other matters. The Ministry responded to a letter from the Tehri Hydro Development Corporation of December 1989 the following July, according conditional clearance. This was done on the assurance that the concerned authorities would furnish detailed action plans in accordance with a prescribed timetable and that these plans would be implemented *pari passu* with the engineering works. (Environment Ministry, July 1990)

Several conditions were imposed. The safety aspects were to be approved by the high level committee constituted for the purpose. Alongside, comprehensive environmental management plans were sought on catchment area treatment based on the recommendations of the Remote Sensing Applications Centre, Lucknow, to identify critically degraded areas in the "very high" and "high" erodibility classifications and to prepare action plans by December 31, 1990; these catchment area treatment plans were to be implemented by the U.P. Government with funds provided by THDC before the end of 1995; the THDC was to get a reputed institution to undertake a socio-economic study of rehabilitation measures to assure the standard of living of those displaced by the Tehri Dam as well as by the Koteshwar dam and those living on the reservoir rim by June and March 1991 respectively; the U.P. Government was required to prepare a plan for command area development by March 1991 and have it implemented by March 1995; flora and fauna studies were to be undertaken in the submergence and other affected areas by May 1991 and an action plan prepared for excution before impoundment; water quality was to be monitored and modelling studies undertaken to prepare an action plan; the THDC was asked to prepare a disaster management plan by March 31, 1991 "in consultation with the likely affected population and the area authorities"; and the Department of Power was to ensure the setting up of a Bhagirathi Basin Management Authority through legislation by March 31, 1991.

These studies and action plans and their implementation were to be scheduled *pari passu* with the engineering works, failing which construction would be halted under the provisions of the Environment (Protection) Act, 1986.

In September 1991, INTACH served a notice on the Environment Ministry to order a halt by the THDC on further work on the Tehri Project as allegedly none of the conditions attached to the environmental clearance had been fulfilled. Sunderlal Bahuguna echoed the demand. The "battle between energy and environment", as it has been termed, continues.

Sedimentation

One of the more forceful arguments after dam safety, used against the Tehri dam is that it will not have the projected life of 163 years, by when alone, THDC says, dead storage will have been completely filled, but only 30 to 40 years. Some alarmists peg the life of the dam even as low as 15-20 years. Sedimentation is expected to be far higher than assumed, as said to be the experience in most or all other large dam projects in the country, and likely to be aggravated by various on-going human interventions.

Mass wasting, deforestation, over-grazing, diversion of marginal lands for agriculture, road-construction and mining activities are among the main causes of erosion in the Himalaya. The flats or benches along the river valleys, and certain poorly-made terraces are unconsolidated and prone to erosion. Alluvial fans deposited by streams at their confluence with major rivers are not cohesive. (Varshney, 1990) All these, together with seismic tremors, landslides, the formation and bursting of debris dams, the exposure of morainal deposits at the terminus of receding glaciers, and periodic cloudbursts are powerful agents of sedimentation. The Kanodia Gad blockage of the Bhagirathi near Uttarkashi in 1978 caused flash floods when the "dam" broke.

All this is well understood and, over time, calculations of sediment formation and transport have undergone refinement. The Tehri project has excluded the unusual Kanodia Gad event of 1978 in assuming sedimentation rates. This gives a siltation rate of 13.49 ha m per annum per 100 square km of catchment which would rise to 16.53 ha/p.a./100 sq km if the 1978 data were included. As against this assumed rate, the actual rate monitored by the Working Group was 13.50 ha m p.a. while the dead storage capacity of the Tehri reservoir has been planned on the basis of an annual siltation rate of 13.97 ha m/100 sq km. The Project provides for a dead storage of 925 m cu m and envisages encroachment of live storage to the extent of 174 m cu m over 100 years. The remaining live storage of 2615 m cu m will gradually decline to 2541 m cu m at the end of 100 years. But, given good catchment area treatment and land use planning, there is no reason why sedimentation rates should not be reduced below these levels to prolong the full operational life of the dam well over the 100 years assumed in relation to its economic viability. The power intake tunnel level will only be silted up after 163 years at current rates of erosion.

A catchment area treatment plan is a necessary part of all modern water resource projects and the THDC is committed to such a programme. The Working Group admits to the "extremely degraded condition" of the Bhagirathi catchment. The Chipko movement opposed felling of trees in Garhwal in order to protect the environment and prevent erosion. Sunderlal Bahuguna coined the Chipko slogan: "What does the forest bear? Soil, water and pure air." Yet he argues that "we are opposing this monstrous Tehri dam because it will devoid (*sic*) the (sugar) cane farmers of western Uttar Pradesh of this precious topsoil which will be trapped in the dam reservoir"! (Bahuguna, June 1992) That is — as so often — having it both ways! The catchment area above the Tehri dam is 7511 sq km, of which 2323 sq km is snowbound. The 4880 sq km catchment below the permanent snowline consists of 1240 sq km of agricultural land, 2400 sq km of reserved forest and 1500 sq km of civil and soyam lands which are managed by the villages and has largely been rendered bare. The U.P. Forest Department has taken up the task of afforestation and densification in 36,000 ha of degraded forest land and had already completed work on 15,000 ha when the Remote Sensing Applications Centre completed its survey of "critically eroded areas" and said these extended over 21,000 ha. This gives a close fit with the Forest Department's numbers. Since then, the area treated has gone up to 22,000 ha. The Environment Ministry is, however, insisting on treatment of a larger area. But the Government has ruled that THDC need only be responsible for treatment of that part of the catchment which lies within its area of influence, namely, the area directly draining into the Tehri reservoir together with areas affected by project roads, colonies, etc. which comes to about 12,000 ha. The THDC is therefore well ahead of its CAT liability. Some "highly erodible areas" have not been treated as the Forest Department finds them inaccessible or under soyam forests over which it has no jurisdiction. CAT is progressing at the rate 2000–2500 ha per annum. The Environment Ministry continues to insist that THDC do more.

Since about 2250 ha of very poor forest lies in the submergence and other Project areas, THDC is undertaking compensatory forestry in double that area in Lalitpur and Jhansi districts where it has been able to locate non-forest lands suited to afforestation. 2800 ha has already been planted through the Forest Department at the cost of the Project. Plantation is being done with a density of 2000 trees per ha in the plains and 1500 trees per ha in the hills where spacing has to be wider. The main problem is to ensure survival, which averages 50 per cent in five years after which natural regeneration takes over. But the Project authorities complain that villagers begin to cut trees at five and six years maturity. Community involvement is obviously necessary.

Flora and Fauna

In its initial survey, the Botanical Survey of India recorded 462 species under 99 families of which 12 might be considered rare and endangered. It felt that botanical gardens or gene banks might be needed to preserve these genetic varieties. It sounded a note of caution about certain aquatic weeds and reeds that might invade the reservoir. A more recent survey by the Botanical Survey has now established that all the 12 rare and possibly endangered species live above the submergence area and are also found elsewhere. So there is no botanical loss. The Zoological Survey of India had likewise done a preliminary reconnaissance which indicated the presence of two important varieties of tree-frog which were slowly vanishing. It expressed concern regarding this as these species are of economic importance as they feed on harmful insects and thereby maintain biological control over timber pests. A second survey is more optimistic. Apart from general afforestation, the Zoological Survey has recommended the planting of fruit trees which provide food to certain animals and of bushy plants along the reservoir rim for the benefit of certain aquatic species. The Forest Department's policy is in any case to plant 20 per cent fruit trees. The bushy varieties will be planted after the reservoir area is clear-felled before impoundment.

The migration of the mahaseer may be affected by the dam which is too high to permit of a fish ladder. The Rishikesh-Chilla hydel barrage already impedes migration runs and needs to be fitted with a fish ladder. The alternative for Tehri is a fish lift or manual transfer in both directions. The matter is to be studied by an expert group. The Working Group had also recommended stocking trout in the upper catchment and smaller streams. This should present no problem.

Dam Safety Issues

Whatever the other criticism levelled against the Tehri Dam, the real point at issue has been safety and the fear of a dam-break should there be a great earthquake of over M 8, as is possible during the life of the dam. The catastrophic consequence of such a disaster have been spelt out in the most graphic terms and there would undoubtedly be a tragedy of the first order were anything like that to happen. The Project authorities and engineers generally are confident beyond any reasonable doubt that the dam is fail-safe and that the grim fears aired are based on mistaken premises, misperceptions and imagination.

The Himalaya was uplifted 40 m years ago when the so-called Indian plate, then joined to what is now Australia and Africa, broke loose and, by a process of continental drift, collided with the Eurasian plate. Plate tectonics suggests that the crustal surface on which we live sits on rafts that ride the molten mass or magma that still boils within the bowels of the Earth to be occasionally spewed out in volcanic eruptions. Himalayan orogeny was spread over geological ages and developed the four ranges dividing the Indo-Gangetic plain that came up with the wash from the Tibetan plateau. The Indian plate continues to thrust and dive under the Eurasian land mass at the rate of five centimetres every year, causing the Himalaya to buckle and shake in the convulsions that we know as earthquakes as the underlying mass slips, tears, faults and deforms to adjust to a new equilibrium, until strains build up to trigger the next seismic event.

Hundreds of tremors occur along the Himalayan seismic belt every year, some too faint to measure. The Richter scale measures the magnitude of earthquakes in terms of the energy released in a progression of M 1 to M 8-plus. No earthquake of M 9 has ever been recorded. The destructive intensity of earthquakes are likewise measured most usually in the Modified Mercalli scale of MM I to XII. It is important not to confuse the two scales, as some commentators on the Tehri controversy have done. The Tehri dam is designed to withstand earthquakes within the broad parameters of M 7.5–8 and MM IX.

The Himalaya represents "one of the most energetic geodynamic processes active today" and accommodates about half the displacement caused by continuing continental convergence through deformation. The remaining strains continue to accumulate until they reach breaking point and are finally released through a catastrophic earthquake of M 8 and above. (Gaur, 1984) The Himalaya has witnessed four such Great Earthquakes in the past century. These were at Shillong in 1897 (M 8.7), in Kangra in 1905 (M 8.5), in Bihar in 1934 (M 8.6) and near the Assam-Burma-China trijunction in 1950 (M 8.6). The Kangra earthquake was felt over 4.16 m sq km and apparently uplifted Dehra Dun by 12.70 cm. Geophysicists believe that there is a major "seismic gap between the rupture zones of the 1905 and 1935 earthquakes (the Central Gap of Khattri *et al*), and that this gap can be expected to be filled in the future with an earthquake of comparable great size. This gap includes the site of the proposed Tehri Dam. A potential slip of considerably more than 10 metres may have been stored up". (Brune, 1991)

Gaur explains that almost the entire length of the Himalaya is characterised by two major intra-crustal boundary thrusts. The Main Central Thrust forms a tectonic boundary between the Great and Lesser Himalaya. While earlier most active, present seismicity appears concentrated along a 30-km wide belt southwest along the Main Boundary Fault, the southern ringe of which is marked by the increasingly active Himalayan Frontal Fault. Noting the proximity of the Tehri dam to the MBF and the apparent hazards therefrom, and the admitted benefits likely to flow from the Project, Gaur posed a set of "meaningful questions" that required to be answered. Though it might well be within Indian engineering capabilities to mitigate the seismic risk to Tehri dam, he proposed a series of research goals to define "a dynamic tectonic model that may be reliable enough for forecasting its near future behaviour" (1984). This should be pursued.

However, the point that became the focus of subsequent debate was his observation that the seismic gap between the Kangra and Bihar earthquakes was long enough (700 km) "to require ruptures by at least three major earthquakes (M 8) from west to east, to release the strain that may have accumulated at the edges. The Tehri region where no major earthquake has occurred since 1828 (M 7.6) may thus have large residual strains already accumulated".

The Working Group to assess the environmental impact of the Tehri dam cited this reference in questioning dam safety. The "environmentally-oriented members questioned the assumed technological infallibility" and in his covering letter, presenting what had become a disputed report, the Chairman in 1986 called for a stoppage of work on the dam. A number of geophysicists and other lay persons joined in with tracts, articles, statements and letters, some of these collated by INTACH (1987). Reference has already been made to the report of the Environmental Appraisal Committee of the Ministry of Environment and Forests (1990) which found the dam unsafe, and the rejection of this finding by the Committee of Secretaries of the Government of India.

Layman's Guide to the Seismic Debate

As the debate progressed, the core of the argument appeared to revolve around the value to be ascribed to the maximum credible earthquake (MCE) for which the Tehri Dam must be designed, with corresponding values of epicentre, focal depth and related parameters which in turn would determine the peak ground acceleration (PGA). The lay person must be forgiven for being bewildered and confused by the passion with which the debate has been conducted in terminology he finds incomprehensible but frightening. A simple layman's guide might therefore be useful, the attempt borrowing heavily on various experts.

An accelograph records the strength of ground motions or vibrations set off by an earthquake of any given magnitude as a fraction of the earth's gravitational acceleration which has a standard value of one. The peak value in the oscillating graph represents the peak ground acceleration. The highest peaks are, however, represented by only a few spikes within the total duration of earthquake shaking and, depending upon the type of structure, may not have any bearing on its design. To capture the severity of a recorded accelogram, engineers use a term called Effective Peak Ground Acceleration which is taken to be equal to 90 per cent of the peak value of the record. Analyses of natural records indicate that EPGA values are generally never more than 50 per cent of the PGA value.

There is some controversy regarding whether the risk level to be considered in choosing the PGA value should be equal to the mean value or the mean plus standard deviation value which would give an 84 per cent dependable value. The International Congress on Large Dams (ICOLD Bulletin No. 72) sets out Guidelines for Selecting Seismic Parameters for Large Dams. This indicates that current design practice for dams tends to rely on the mean rather than on extreme PGA values because of compensating conservatism applied elsewhere. Thus, for example, with regard to spectral ordinates, engineers choose a synthetic accelogram which is more severe than the natural accelogram. Everybody, the earth included, responds to any vibration in a given pattern. The time taken to complete a full cycle of free vibration is called the natural period of vibration of that structure. If the period of vibration is close to the natural period of vibration of the structure, the latter may be amplified, resulting in its failure. This phenomenon is known as resonance. Thus, when a body of foot soldiers crosses a bridge, the troopers are advised to march out of step lest the rhythm of their otherwise uniform tread coincide with the oscillation of the structure and bring it down. Likewise, the voice of a singer can have a shattering effect if the oscillations of his or her voice resonates with the natural response period of a wine glass.

Thus, the behaviour of a structure in response to a particular type of vibration loading will depend upon its natural or fundamental period. If the period of the structure and that of the vibrations are in close proximity, the resultant resonance can have undesirable results. Earthquake energy is transmitted in short period waves, but a high dam has a higher natural period and is, therefore, inherently stable. A concrete dam, being rigid, would tend to snap more easily than an earth and rockfill dam like Tehri whose huge wedge-like mass with a 1125 m wide base and gentle slopes would invest it with greater inertial resistance to shaking. The energy damping quality of rockfill material also imparts greater resilience to the structure. In short, a high, rockfill dam has inherent properties of that make for safety.

In designing a dam in a seismic-prone region, note is taken of all the proximate faults and tears around the site along which displacement might take place in the event of an earthquake. Each fault within the relevant seismic province of the dam will have an upper bound to what could be a Maximum Credible Earthquake (MCE). But the greatest of these values within this tectonic envelop is the critical MCE. Since it is now established that there is no fault in the river bed beneath the Tehri dam site, any earthquake is likely to have its epicentre some distance away and at a focal depth of 15 to 40 kms as actual measurements of past seismic events indicate. Thus the ground motion from the critical MCE will be attenuated by the dampening effect of the strata through which it traverses before it reaches the dam. This value is taken as the Maximum Design Earthquake (MDE). Within these parameters, an analysis is conducted of the degree of damage or deformation the dam might suffer within permissible limits in the event of an earthquake, it being assumed that risk of an actual dam-break is precluded by the design. Appropriate defensive measures are then incorporated in the design to ensure against overtopping by wave action, liquefaction of the foundation or fill material under intense pore pressure.

The Tehri Dam is sited in Seismic Zone IV in the Seismic Zoning Map of India incorporated in the ISI Code on Indian Standard Criteria for Earthquake Resistant Design of Structures (IS: 1893-1975, Third Revision). The Standing Committee for Evaluation of Seismic Coefficients for River Valley Projects of the Government of India recommended in 1984 a seismic coefficient of 0.12 g for what is termed the pseudo-static design of the Tehri Dam. The Department of Earthquake Engineering, Roorkee University, in 1983 recommended an effective peak ground accelaration of 0.25 g corresponding to an MCE for the dam, taking into account various faults mapped in the region (Working Group, 1986). The subject of debate thereafter has been whether a PGA of 0.25 g is adequate, which the critics strongly contest, and whether a MCE of M 7, for which the dam has been designed, is realistic, or as the critics again argue, is too low and should be M 8 or more.

The critics' point of view was set out in INTACH's plaint before the Supreme Court (1993). This stated that until 1990, THDC maintained an MCE value of M 7 with its epicentre 27 km from the dam and with a focal depth of 20 km, generating a PGA of 0.25 g. However, despite subsequently assuming an MCE of M 8.5, a 300-times increase in energy release, the High Level Committee concluded that Effective PGA would still not exceed 0.25 g. This was said to be untenable as there has never been a recorded earthquake of M 8 and above whose Effective PGA has been as low as 0.25 whereas lesser earthquakes have registered an effective PGA equal to or greater than 1 g. It is also incomprehensible that two earthquakes of M 7 and M 8.5 will generate the same level of ground acceleration. The devastating impact of the M 6.1 Uttarkashi earthquake in October 1991 was cited as sufficient indication of the far greater havoc likely to be caused by an M 8.5 event.

The High Level Committee set up by the Department of Mines had earlier answered these doubts (April 1990). In order to instil public confidence in the Tehri design, as further endorsed by the Soviet consultants, it was decided to test it under a worst-case scenario. This assumed an M 8.5 earthquake right under the dam at a focal depth of 15 km. Even so, the test yielded an Effective PGA of no more than 0.22, or less than the value adopted in the design. The dam slopes were found stable and the deformation or settlement of 0.5 m was more than accommodated by the provision of a 9.5 m freeboard. The shell material proposed to be used in the construction of the dam when subjected to an Effective PGA of 0.25 g did not show signs of liquefaction. The HLC did not find any danger of reservoir-induced seismicity (RIS) as a result of filling of the reservoir and the load this might engender, nor any lubricating effect from possible water seepage that could cause a slip. Seismicity induced by a large artificial reservoir can only act as a trigger to initiate a natural earthquake that would have occurred otherwise, but cannot increase its size. (Supreme Court, November 1990)

Dr. Vinod Gaur's dissenting opinion was reviewed in a supplementary report of the HLC and his further reservation was responded to by Dr. Jai Krishna as earlier recounted. It is Dr. Jai Krishna's opinion that "the intensity of ground motion does not go on increasing indefinitely with the magnitude of an earthquake. Bigger magnitude earthquakes exert strong ground motion over a bigger area without their intensity increasing, since there is an upper limit to the resilience of a material and its release of energy per unit volume that ruptures at a time to cause an earthquake. Thus for a particular site, the magnitude of an expected earthquake of 7 or 7.5 or 8 or 8.5 means approximately the same maximum intensity of ground motion and the same response of structures. Although, in terms of total energy released, the ratio between magnitude 8.5 and 7 may be about 175, the intensity of damage at any particular location will be of the same order". The area of heavy damage will expand with great magnitudes. Likewise, "a strong peak (in an accelogram) can be larger than 1 g in a moderate shock of M 6 and no peak may be greater than 0.5 g in a shock of M 8. Thus peak ground acceleration by itself has no material significance .... Taller dams have a smaller response than small dams and are therefore safer against earthquakes". And finally: "Since 'Magnitude' is a logarithmic number, the maximum credible earthquake should be chosen carefully. An increase of 0.2 in magnitude provides a margin of safety of 2 when measured in terms of energy released. Thus, large increases in fixing the size of (the) expected biggest earthquake leads to over-provision". (Krishna, Jai, December 1991)

Another interesting study records earthquakes that have a PGA exceeding 1 g near the causative fault but show a considerable decline in PGA values 15 km away. The relative values are set out in the following table.

The THDC is convinced that the Tehri dam design, prepared after the widest possible consultation with various expert and apex bodies and institutions, is safe. This has been independently confirmed by Soviet/Russian experts who have a great deal of experience in building high rockfill dams in areas of very high seismicity and have subjected the Tehri design to state-of-the-art two-dimensional dynamic analysis. They have confirmed the basic design and dimensions of the dam while suggesting minor

Table : Earthquakes with Peak Ground Acceleration (P.G.A.) Exceeding '1 g'

Earthquake	P.G.A. Value> '1 g'		P.G.A. Value at 15 km away from Causative Fault	
	Distance from causative fault in km.	*P.G.A. as % of 'g'*	*Distance from cautive fault in km.*	*P.G.A. as % of 'g'*
San Fernando Cal, U.S.A.	3.2	1.25	15.4	0.225
Imperial Valley Earthquake	5.6	1.75	15.0	0.163
Victoria Earthquake	7.3	1.0	15.0	0.154
Nahanni Earthquake	9.0	< 2 g.	16.0	0.190
Coyote Creek Dam Earthquake	3.2	1.29	14.4	0.40
Gazli Earthquake	3.5	1.35	No data available.	
Tabas Earthquake	3.0	0.94	17.0	0.39

Source : Tehri Hydro Development Corporation, New Delhi. 1993.

changes in the shape of inner core filling. Damage to a structure comes from localised energy release unrelated to the magnitude of the earthquake at its source. While critics have set much store by Prof. Brune's formula, Indian earthquake engineers maintain that this is only one of several empirical formulae used for calculating attenuation of energy. Analysis of the Tehri design on the parameters of the accelograph of the Gazli earthquake, which had a PGA in excess of 1 g, was desired by critics. This was done by applying an acceleration of 1.36 g vertical and 0.72 g horizontal, both acting simultaneously. The dam was found safe. (THDC Directors' Report, 1992) Objections to the narrow canyon effect at Tehri on the Gazli test are countered by THDC experts. The argument that the Tehri design should be further subjected to 3-D dynamic analysis is answered by the plea that this procedure is still being developed and has not been fully applied anywhere as yet except on a simplified basis.

The Uttarkashi earthquake triggered fresh questioning about the safety of Tehri dam. It was argued that if what was initially thought to have been a M 6.1 earthquake had wrought such havoc, what might not happen were the threatened 'big one' of

M 8-plus to visit the region? Most of the damage, however, was to poorly constructed dwellings. The magnitude of the earthquake, determined to be M 7.2, was well within the design parameters of the Tehri dam whose appurtenant works were totally unaffected, contrary to rumours of some cracks. Indeed, though the epicentre was in the vicinity of the Main Central Thrust at a focal depth of 10-15 km and very near Maneri Bhali-I, that 45 m-high hydel dam suffered no distress of any kind. A PGA of 0.3 g at Uttarkashi was damped down to 0.07 g at Tehri. (Chandrasekaran and Das, December 1991)

Although the Uttarkashi earthquake was irrelevant to any issue of dam safety at Tehri, public agitation and a fast unto death by Sunderlal Bahuguna persuaded the Prime Minister to appoint yet one more committee to resolve the dam safety controversy once and for all. A whole lot of distinguished experts participated and in mid-1992 concluded that, having done the dynamic analysis of the Tehri dam on the parameters of the Gazli earthquake and reviewed all other issues, the Tehri dam was safe. Among the defensive measures taken to protect the dam are compaction of the rockfill and earth material to a density approximating concrete to preclude any risk of liquefaction; prevention of any ill-effects of magnification of earthquake acceleration at the top of the dam by widening the crest to 20 m, and further to 25 m at the abutments; provision of a liberal freeboard of 9.5 m to insure against any slump or settlement or risk of wave action; providing for a fine sand layer on the upstream face of the dam which, in the unlikely event of a crack developing in the core, will be washed into such cracks as a self-healing process; installation of a downstream filter to prevent any erosive migration of fine particles in the core should it develop a crack and admit water; and designing flatter dam slopes than those constructed in high-seismicity areas elsewhere. Finally, as a matter of abundant caution, the dam section has been checked for a PGA of 0.5 and an earthquake in excess of M 8. Other large dams constructed in the Himalaya such as Bhakra, Pong and Ramganga in India and Tarbela and Mangla in Pakistan have shown no sign of reservoir-induced seismicity which, in any case, would be of a value far less than the MCE for which Tehri dam has been designed. The dam will not break. (THDC, 1992)

Further controversy has not been stilled. INTACH conducted a workshop in New Delhi in January 1993. There, Bruce Bolt, Professor of Seismology at the University of California suggested that arriving at a suitable number for a Maximum Credible Earthquake tends to involve subjective factors regarding what is credible. Therefore, the U.S. Division of Dam Safety is being sought to be persuaded to adopt the concept of a Safety Evaluation Earthquake (SEE), this being "the earthquake whose occurrence is judged to be sufficiently unlikely so that it can represent the maximum earthquake that should be considered in evaluating the safety of the facility being planned". Bolt believes it important to assess each stage of the parameters in a conservative way, taking into account the consequences of failure. This situation, he adds, "raises the basic question of acceptable risk . . . . There is always a trade-off between life safety and cost and between life safety and economic benefit to the population". (Bolt 1993)

An Indian Professor, R. N. Iyengar, echoed this view. After evaluating the Tehri design parameters he concluded: ". . . The dam may be safe; but it does not appear to be sufficiently safe at present. Safety is a concept that all of us understand but is difficult to quantify. . . . The risk of failure due to earthquakes can be lessened to small levels. But such an exercise will invariably be very expensive. It is for the planners calmly to decide how much the country can spend in relation to how much risk the public can tolerate in the postulated scenario of a seismic failure of the Tehri dam in the culturally important Himalaya. This is not just the engineer's problem. The planners and the public should actively participate in the decision-making process." (Iyengar, January 1993)

Prof. Iyengar's concern about the canyon effect of a narrowly confined gorge such as at Tehri was not ignored in the design. Following the INTACH workshop in January 1993, the National Committee on Seismic Design Parameters met to discuss some of the issues raised. Prof. Iyengar also attended this meeting by invitation. He appears to have conceded that as a 3-D analysis of the Tehri design might not be immediately necessary or practical, the time period of the dam (calculated at 1.8 to 2.5 seconds under 2-D analysis) should be assumed to be 0.8 to 1 second (as would more likely be the case under 3-D analysis), and the design tested accordingly. He was informed that the dam was indeed tested under an assumed 0.8 seconds time period by Soviet engineers and found to be perfectly safe. In sum, THDC officials believe that Prof. Iyengar's points are adequately met. INTACH had in fact sought a clarification from Prof. Iyengar some months after the workshop and forwarded his response to the Prime Minister in August seeking a further review of the Tehri dam. (Bhagirath ki Pukar, August 1993) The Ministry of Power must no doubt have informed the Prime Minister of the discussions held by the National Committee on Seismic Design Parameters.

Rim Stability

Even given the dam is safe, anxieties have sometimes been expressed about slope failure as the reservoir rim is unstable and hillsides could slide into the lake, generating dangerous waves that could overtop and destroy the dam. With the normal reservoir level at 830 m and the dead storage level at 740 m, the water level in the lake could fluctuate by up to 90 m between October and June each year. It was at one time feared that impounding could result in reworking of some slopes with bank erosion in soft rock areas and sliding or slipping in hard rock areas. The Working Group sought a geomorphological study by the Wadia Institute for Himalayan Geology whose somewhat pessimistic findings were challenged both by the Geological Survey and Roorkee University. The latter made a detailed investigation and found the rim slopes generally stable. There could be slides of surface overburden. Local slips from wave action and wetting were expected to attain a new angle of repose on submergence. But no area was observed where a large scale hill-slide might generate a destructive wave that could endanger the dam. A slide of the kind that occurred in October 1963 upstream of the Vioant Dam in Italy, creating a wave which topped the dam killing 1900 people in the valley below, is ruled out at Tehri. The circumstances and conditions in the two cases do not match to warrant a close comparison.

The stability of villages above the reservoir rim was also carefully examined by Roorkee University and found satisfactory. The condition of these villages was rechecked after the Uttarkashi earthquake and found to be all right.

Yet another doubt was raised with regard to sedimentation. This was that siltation could cause a deltaic barrier to form at the far end of the reservoir and that such aggradation might result in backwater flooding of Uttarkashi a little upstream of that point. This too is not seen to be a problem.

A command area development plan is being prepared by the U.P. Irrigation Department and responsibility for its execution has been taken by the State Government. Likewise, a catchment area authority for the Bhagirathi-Alaknanda catchment is proposed to be set up by the U.P. administration to prepare suitable development plans for the region. The State Government wanted to do this by executive order, but the Environment Ministry is insistent on the setting up of a statutory body. Water chemistry has been examined and no problem is seen as likely to emerge on this count.

A Disaster Management Plan has been prepared by the Centre for Development Studies as required and submitted to the Ministry of Agriculture which is the nodal agency for dealing with natural disasters. The THDC says its study covers flood risks and possible slumping or deformation with pre-warning systems based on compete instrumentation of the dam to detect distress. Communication systems have been devised for relaying flood warnings downstream. The Environment Ministry is, however, asking for a dam-break study which THDC is reluctant to do as it feels that, given the highly exaggerated fears that have been spread regarding the safety of the dam, even a theoretical dam-break exercise, never so far undertaken in India, could cause needless panic. The THDC disaster management plan has no doubt been done in accordance with the CWC's Dam Safety Procedures which has a separate chapter on Emergency Preparedness for both natural and manmade disasters, encompassing disaster prevention and disaster preparedness. The procedures include disaster evaluation, likely effects on life and property based on inundation mapping, vulnerability analysis and hazard area mapping, review of anti-disaster shelter adequacy, medical aid, transportation, emergency food supplies and warning systems. Certain dam-break studies based on computer packages regarding floods are available. Arrangements for public safety can be devised thereafter with public participation. (CWC, 1986)

It would nonetheless be useful to dispel grosser exaggerations such as the fanciful statements made even by responsible individuals and agencies that a dam-break is not merely likely but, should that happen, would wash away everything up to Kanpur or even Calcutta. Even assuming the reservoir is full in the unlikely event of a dam-break, 2615 m cu m of live storage would be dissipated in a little over 50 square km with inundation to a depth of one metre up to Hardwar where the valley opens out into the plains. Part of the Doon Valley would be submerged. Though the possibility of anything like this happening are at best extremely remote, inundation mapping of this kind could help instil a measure of confidence at least among the better educated population. The Central Board of Irrigation and Power did conduct a conference on Research Needs in Dam Safety in 1991 where a dam-break flood hazard analysis was presented. (Reuter, December 1991) British Columbia Hydro (Canada) has also been evolving improved dam safety programmes including the preparation of Emergency Preparedness Plans for each dam. The EPP's are regularly tested and a relevant operational exercise involved a simulated earthquake dam breach. BC Hydro's Corporate Emergency Centre in Vancouver was activated and 23 agencies participated in the community emergency operating centre set up near the dam. Public acceptance was reported good. (Nielsen, March 1993)

As far as the Government's environment clearance conditionalities are concerned, with the delay in the Tehri project schedule, first on account of the seismic controversy and then the financial uncertainties following the break up of the Soviet Union, the various environment studies, surveys and action plans submitted to the Environment Ministry and their implementation is not merely running *pari passu* with the engineering works but is now well ahead of that in all cases. Rehabilitation too is being synchronised with the schedule of works and impoundment.

Resettlement and Rehabilitation

The most traumatic aspect of any water resource development project is displacement from lands going under submergence or acquired for other purposes. All displaced persons need to be treated with compassion and generosity so that not just resettlement but rehabilitation, a concept of much deeper meaning, takes place and those evicted are no worse but in fact better off than previously at the end of the exercise. Like certain other projects in India, the Tehri dam complex has been inflicted with delay. This has prolonged the agony of those facing or liable to face displacement, the very uncertainty adding to the emotional burden. Foreknowledge of submersion has also inevitably resulted in a slow down of normal development activity and even maintenance of facilities which, when the project is delayed, causes hardship. Delay has also led to lands being acquired and affected families being compensated at different times. With an improvement in the scope and terms of compensation with the passage of time, this has meant differential compensation and a sense of grievance on the part of the early awardees who feel discriminated against merely because they cooperated sooner than others in some cases. Since rehabilitation is a psychological rather than a material concept, speed and certainty are supremely important.

The Tehri Project will submerge Tehri town and 112 villages, 69 of these partially. Some villages have been or will also be acquired for New Tehri Township and for other project-related purposes. In all some 9290 rural and 4551 urban families are affected or a total population of nearly 70,000 persons. Those being displaced from Tehri town are being rehabilitated in New Tehri Township (NTT) located on a ridge above the reservoir. The rural displaced families are being resettled in various sites in Dehra Dun and Hardwar districts.

The resettlement and rehabilitation (R & R) policy was determined in consultation with authorised representatives of the oustees as they are unfortunately termed in official parlance. The average rural holding of the displaced agriculturists in the area acquired was 0.45 ha but they are being guaranteed a minimum of 0.8 ha (two acres). This minimal compensation has been made applicable to agricultural labourers as well. Major sons are excluded, though it is now being agitated that they too should be treated as independent families for purposes of R & R as under the Sardar Sarovar package in Gujarat. The lands offered are developed plots with irrigation facilities. The minimum house compensation is Rs. 40,000 but the average works out to Rs. 80,000 plus an ex-gratia payment of Rs. 50,000 per dwelling. Cement, GI sheets and steel are being made available by the THDC for housing construction. Overhead tanks are being provided, with water service mains and house-to-house connections for drinking water supplies. This has been a great boon. Standard civic facilities such as roads and education and health amenities have been provided, with educational institutions up to the higher secondary level. All fields are connected by chak roads. The residential plots of 200 sq m are being provided at actual cost. The R & R package is likely to cost THDC around Rs. 3.32 lakhs per family, including a resettlement grant and a transportation allowance.

Those families preferring to take cash compensation and settle on their own are entitled to do so. They are being paid Rs. 12,000 per acre for irrigated land, and Rs. 8000 and Rs. 4000 for Class I and Class II unirrigated lands, subject to a minimum of Rs. 40,000 plus 30 per cent of the land cost as solatium. (THDC, 1993)

New Tehri Town has come up and promises to become a fine hill station overlooking the proposed Tehri lake. Each urban displaced family of old Tehri town will get a residential and/or commercial plot varying from 150 to 300 sq m size on a slab rate related to area from Rs. 5 per sq metre up to 150 sq m and Rs. 250 per sq m in the highest slab. Each displaced family will get a minimum housing plot of 60 sq m and a displacement grant for household effects and commercial goods at given rates. Compensation for the houses acquired in old Tehri is being paid at the prevailing market rate at the time of acquisition with an ex-gratia payment up to a maximum of Rs. 50,000. No house compensation is less than Rs. 20,000. All religious and cultural institutions are being re-established on a far more generous scale than before.

About 80 per cent of all district offices and government employees had shifted to NTT by mid-1993. Private plots have been allotted but construction has yet to commence in some cases as the owners have residences or businesses elsewhere and are therefore in no hurry to build. Garhwal University has been provided a spacious campus, a Girls College has been built as well as an ITI and a Training Centre. A fine hospital, bus stand, municipal building, and places of worship are planned to replace whatever is going under. Tree plantation and construction of footpaths are in progress. New Tehri Town is being developed as a symbol.

Economically weaker sections with a compensation not exceeding Rs. 20,000 are entitled to surrender this in exchange for a two-room house on a 60-metre plot. THDC is constructing 300 such houses and 120 families have already moved. The commercial premises are being built with residential accommodation above them.

The rural displaced fall into two categories. The earliest to be displaced are those who will be submerged by the construction of the coffer dam for diverting the river. The coffer dam, which is later to be incorporated as an integral part of the main dam, had risen 15 m above the river bed at the end of the 1992-93 construction season. The coffer dam villages (about 1500 families) have been resettled in five locations in the Doon Valley, some of which was prime sal forest that was, regrettably, clear -felled during the 1970s before the Forest Conservation Act came into force. Another 500 families have opted for cash compensation and have resettled on their own. About 23 families remained to be settled in mid-1993. They are seeking half-acre plots in Dehra Dun where the land cost is Rs. 2.5 lakhs per acre, with services costing another lakh of rupees. They prefer this to R & R in the neighbouring Saharanpur district. This shows a change from the old money order economy of the Hills, where men migrate and the women are left to manage the farms, to non-agricultural, urban occupations with the whole family moving and living as a unit.

The second, and major, component of rural displacement will come with the filling of the Tehri reservoir. 2800 families are involved. The stipulated two-acre plots are so far available for 500 families and 2300 more families remained to be provided for in mid-1993. About 500 families may opt for cash compensation of around Rs. 1 lakh each in lieu of land. The residual 1800 families will require about 7000 acres of land inclusive of farms, residential plots and service area. The villagers are being consulted about moving to Hardwar or Saharanpur districts. Persons in partially-affected villages will receive compensation and will be given first preference in farming the 450 ha or so of land that might be available for draw-down cultivation along the upper reaches of the reservoir. Complaints of delays in R & R and grant of proprietory rights over the land are partly true. Organisational changes did not help. R & R was initially under the U.P. Irrigation Department. This was then transferred to the autonomous Tehri Dam Organisation set up in Dehra Dun. Turf problems arose and R & R files were not transferred. Moreover, TDO officials lacked the revenue and magisterial powers that were vested in government officials earlier dealing with the subject. Hopefully, these problems have by now been largely overcome and the THDC is pursuing the matter more effectively.

There has been some criticism about the scale of R & R in relation to the Sardar Sarovar package and otherwise, and the manner of its implementation from time to time. The Environment Ministry had sought an independent evaluation and this was done by the Administrative Staff College, Hyderabad. Its findings are revealing. The study says "that the beneficiaries of THDC under the rehabilitation programme are undoubtedly better off than what they were earlier. THDC is making sincere efforts to rehabilitate the families and construct the irrigation systems. However, the programme so far does not appear to be satisfactory because of the resistance faced from the local population, ill-informed youth (and) non-government organisations, as well as inadequate flow of funds. If these hurdles are removed, Tehri project is bound to generate a bright future for the area and its inhabitants". (Administrative Staff College, December 1992)

The ASC evaluation is related to the coffer dam oustees. It indicates that none is engaged in agricultural labour after rehabilitation, which implies that all landless agricultural workers have received land of their own. There is no conflict with the host population. The value of assets of those rehabilitated has risen from Rs. 46,371 per dwelling unit to Rs. 1,31,343 for the new concrete houses. Income from paddy and wheat have increased by 131 per cent and 119 per cent respectively and overall farm production, including that of vegetables, has risen. Agricultural incomes have gone up. Whereas prior to R & R only 9.3 per cent of the households had an annual income of over Rs. 10,000, this percentage has risen to 41. Fewer families keep animals but those that do earn more from dairying. The income from non-agricultural activities has increased by almost 40 per cent. The total average household income, per capita incomes and average family expenditure have all registered significant growth.

The Study notes that dissatisfied beneficiaries are by and large those who received compensation some years ago when land values were lower. Some respondents have sought a "disturbance allowance". A survey revealed incomplete information about submergence, which suggests the need for better information systems. Many families showed strong attachment to their place of birth. The civic amenities "are of superior quality" than in the original villages. This is especially found to be so with regard to education, health and drinking water. Facilities for these in the former locations were inadequate and distant and entailed considerable walking. The availability of piped water in homes "particularly benefits womenfolk as the time saved could be gainfully employed, besides reduction in physical strain". Women have also greatly benefitted as they no longer have to scour the forest for fuel and fodder. LPG connections are available. The houses are better constructed, are nearer the farms, and are bigger and more hygienic. P & T, power and road transport facilities are available. The quality of land is better and the irrigation facilities more dependable. The agricultural holdings are also consolidated and not fragmented as before. Opportunities for wage and self-employment have also greatly improved. In sum, "the quality of life of the rehabilitated family is far above and better than what was the situation before rehabilitation".

New Phasing and Costing

The dam safety controversy and uncertainties regarding Soviet/Russian financing have meant time overruns and cost escalation. Since blasting was banned until the seismic issue could be resolved and there was a shortage of funds, THDC decided to concentrate on New Tehri Town construction and rehabilitation and environmental works during 1991-93, setting aside engineering works until the dust settled. Between 1986 and 1991, diversion tunnels had been completed and the coffer dam raised to 10 metres above bed level. The Soviet committment of a financial package of 1000 million roubles (around Rs. 2000–3000 crores, depending on the rupee-rouble rate of exchange) having virtually broken down, it was found necessary to re-phase and re-schedule the project. Since the Government of UP and the Government of India were committed to funding construction of the main Dam and all environmental works and R & R, these could continue and will, as before, be funded from domestic sources. It was Power Houses Stage I and II and the Koteshwar dam and power house that were to be financed by the Soviet Union, together with a 765 KV transmission line from Delhi to a point near Meerut. Funds for these segments must now be found.

In order to facilitate both construction and funding, the Government has approved a new phasing of the Project. Phase I, to be financed from internal resources, will consist of the main Tehri dam and Stage I power house (1000 MW), together with environmental and R & R components. Phase II will consist of the Koteshwar dam and power house (400 MW). Having thus created a balancing reservoir, Phase III will comprise the second power house at Tehri Dam with reversible turbines for pumped storage (1000 MW). This will be taken up as and when finances can be mobilised.

As adequate resources are not available for Phase I funding in the Eighth Plan, this will spill over into the Ninth Plan, with irrigation commencing at the end of 1997 following partial filling of the reservoir. The four 250 MW turbines will be commissioned in pairs in 1997-98 and 1998-99.

The cost of Phase I is now estimated at Rs. 2815 crores of which Rs. 717 crores had been spent by March 1993; that of Phase II is Rs. 725 crores of which Rs. 15 crores had already been spent; and that of Phase III is Rs. 1224 crores of which Rs. 25 crores had been spent by 1993 to complete such works as must be done alongside construction of the main dam. This puts the total Project cost at Rs. 4764 crores, of which Rs. 756 crores had been spent by the end of March 1993. With environmental clearance and Public Investment Board approval for this phasing in hand , THDC plans to approach the Cabinet Committee on Economic Affairs to clear Phase I with a residual outlay of around Rs. 2100 crores.

As far as Phases II and III are concerned a Soviet-Brazilian-Swiss consortium had shown some interest earlier. Now a Ukranian-Asea Brown Boveri (Swiss) consortium has made an offer of suppliers credit, possibly with Swedish partnership, on the basis of a down payment, a further payment on delivery and a final 50 per cent payment in instalments over three years at 8.5 per cent interest. The THDC hopes to get clearance for the entire complex subject to approval of the financial arrangements for Phases II and III, whenever settled. The cost of the Phase I environmental action plan is put at Rs. 435 crores of which Rs. 253 crores has already been spent, placing the Project ahead of the *pari passu* requirement. Rs. 268 crores is estimated to be spent on R & R.

The THDC has a sanctioned budget of Rs. 100 crores in 1993-94 and could take up works valued at up to Rs. 150 crores if more funds are forthcoming. It would in any case like to complete the coffer dam section by June 1994. It will be better placed to go to the market for loans or debentures once it begins generating power. Its negotiations with foreign suppliers will also only pick up after Cabinet clearance which will remove any lurking ambiguity about the future of the entire complex, including Phases II and III. The Tehri Project as a whole will produce 4238 m units of energy annually and will firm up hydel output at the downstream Chilla power house (by about 200 m units) and on the mini-hydel stations on canal falls lower down. The unit cost of power is likely to be 77 P/kWh for Phase I, 56 P for Phase II, and Rs. 1.30 for Phase III, compared with a current unit cost of around Rs. 3 in the case of gas turbines and Rs. 1.50 to Rs. 1.70 in the case of the UP State Electricity Board. With 80 per cent of the Tehri Project cost debited to power and 20 per cent to energy, the cost of power per installed MW will be under Rs. 3 crores as against a corresponding figure of Rs. 3–3.5 crores per MW in the case of present-day thermal stations. But then the generation cost of hydel is far cheaper as no fuel (coal) or mining and transport costs are involved.

The Phase I power house will operate at 40 per cent load factor as it will cater to/base load. The Phase III station will only supply peaking power of which there is a 20–24 per cent shortage at present in the Northern grid which also has a rather unfavourable hydro-thermal mix of 28 : 72 that could do with correction for the stability and profitability of the system.

Other benefits from the Tehri complex will be an improvement of roads and communication which will give a fillip to horticulture and vegetable gardening, animal husbandry, tourism and industrial and self-employment opportunities with the creation of superior educational and training facilities. Reservoir fishery should flourish. Catchment area treatment and afforestation will improve a degraded environment. The infrastructural and income-generation potential of the complex will trigger wider area development.

The Tehri project has been faulted on a number of techno-economic grounds including its financial unviability, short life (owing to sedimentation), inherent seismic hazard and so on. In an elaborate exercise that in effect inflated costs and reduced benefits, Vijay Parañjpye some years ago came to the conclusion that the benefit-cost ratio for irrigation would be 1.28 and not 3.68 as estimated by the Project authorities. Likewise, the revenue return on the power component was estimated by him to be 6.89 per cent and not 11.52 per cent. His overall conclusion was that the B : C ratio for the Project would be 0.56 : 1 which would render it non-viable and otherwise unjustified in view of the risk of dam-collapse. (Paranjpye, 1988)

These arguments were reviewed and answered by a distinguished water economist who noted the speculative nature of benefit-cost analyses and said that if the project authorities had oversold the Tehri dam, "Paranjpye cannot escape a like charge in reverse, that is, killing (what is) probably a viable project". He referred to the chronic under-pricing of water and electricity in India and urged that, as a critical infrastructural service, power should be priced at its marginal value product in the economy. He found an underassessment in the irrigation benefits assumed by the Project authorities and further scaled down by Paranjpye, and placed great emphasis on something quite ignored by Paranjpye, namely, the very considerable indirect impact on farm output of surface irrigation through groundwater recharge. All things considered, he found the doubts cast on the economic viability of the Tehri project to be not well founded in facts. (Dhawan, August 1991)

The B : C ratio and rates of return of the Tehri Project have undergone change over the past couple of years with changing circumstances. Nevertheless, the THDC is confident that the B : C ratio for irrigation will not be less than 2.5, well above the Planning Commission's threshold figure of 1.5, and that the return on power will remain attractive in itself and in so far as the Tehri complex also improves the overall hydro-thermal ratio in the UP-Northern grid, provides valuable peaking power, firms up downstream hydel generation and stabilises the grid. The proposed 300 cusec supply to Delhi for drinking water will have an increasingly high value, and may later even need to be enhanced to meet the mounting shortage of water in the national Capital.

Upper Ganga Basin Development

The battle over Tehri has been long and hard. But this is not the only project coming up in the Upper Ganga and Yamuna Valleys. A series of projects on the Yamuna have harnessed about 475 MW of its potential. The larger Kishau and Lakhwar-Vyasi projects on the Yamuna-Tons have been shelved for the moment on account of lack of funds. On the Upper Ganga, above the Rishikesh-Chilla project (144 MW), Maneri-Bhali I (93 MW) is operational again after some trouble with silica corrosion of its turbine blades. Work on Maneri Bhali-II (248 MW) just downstream of that has been suspended on account of financial constraints which have also brought work on the Srinagar project on the Alaknanda to a standstill. Environmental objections to the (480 MW) Vishnuprayag project on the Alaknanda and its Bhuyandar hydel link have been stalled on account of fears that tunnelling could trigger landslides at Joshimath, the roadhead before Badrinath, and cause environmental damage to the Valley of Flowers. A private entrepreneur has however now expressed interest in taking up this project. A whole series of schemes, mostly run-of-river, but some storage dams like Kotlibehl (1000 MW) on the main stem of the Ganga between Rishikesh and Devprayag, are planned to utilise the available gradients and sites in Garhwal which have a potential of around 10,000 MW. Not all of these may fructify, as important pilgrim centres and cultural sites have to be safeguarded and the environment protected. Micro and mini hydel possibilities abound.

Tehri dam is only a beginning. It is of course vitally important to develop a viable basin plan based on techno-economic, environmental and socio-cultural considerations, fitting each new project into a larger mosaic of overall development. This will alarm some, though run-of-river projects pose no seismic danger, and dismay others who would like the Ganga and Garhwal to remain unsullied. It would be desirable to maintain some tributary streams in their pristine condition and to preserve selected wilderness areas. But the people of Garhwal too want development and opportunity and there is no reason why they should be denied these. Far from Tehri dam spelling doom, water resource development, sensibly planned, with due caution and sound land use planning, could green the Himalaya. The attempt to load the Tehri dam with moral and financial responsibility for treating the entire Bhagirathi catchment is quite unrealistic. But each project, within its sphere of influence, would need to take up catchment treatment and watershed management programmes that could cumulatively be quite transforming. A new hill agriculture, moving away from grain to more ecologically-suited cropping patterns and stall-fed dairying is feasible, now that roads and communications ensure food security to the remotest Himalayan hamlets. Let the irrigated plains feed the hills and let a cess on hydel generation be employed to fund such programmes.

The human erosion caused by migration from hill to plain, especially of men, leaving women behind, must be reversed and the long neglect of the hills and their steady environmental degradation — from sheer pressure of population and livestock more than by any other factor — must be ended. There are instances in the Garhwal Himalaya where more than three-fourths of the families migrated out of a village and an equal proportion of cropped land was abandoned during the period 1972–88. It is for salaried jobs for which males migrate but ownership of native land is retained. Management of land in such situations becomes the exclusive responsibility of women. Enormous expenditure of energy and time in collecting fuelwood, fodder and drinking water have made life miserable specifically of rural women". (G.B. Pant Institute, 1992) Who has calculated the B : C ratio of this gender inequity? Environmental fundamentalism can only injure both Himalayan welfare and ecology. Extravagant statements libelling the Tehri project a "scandal" and a "planned disaster", "an assault on the Hindus' most sacred symbol", and the assertion that "there is just not enough water" in the Bhagirathi to provide the planned power and irrigation (Vandana Shiva, April 1992) border on hysteria. So also the hyperbole about the "most heinous crime" of "choking the sources of the Ganga" and questioning the moral authority of those who would "destroy an entire river in pursuance of some insane and totally wicked notion of development". (Nandy, Pritish, April 1992) We are further told that "several hundred thousand fishermen will be thrown out of employment as the movement of fish will be obstructed" by Himalayan dams; that the "four gifts of the Tehri dam will be "atrocities, displacement, corruption and genocide". (Sunderlal Bahuguna, July, 1993) A tidal wave of nostalgia and emotion threatens to sweep away reason. The danger of eco-extremism is graphically illustrated by the fact that rigid centralisation in implementation of the Forest Conservation Act until recently, has engendered a *jungle kato andolan* (fell the forest movement) in Garhwal's Chipko country. (Agarwal and Narain, March 1991)

The seismic debate, scientifically conducted, was in order. But it got derailed once people were labelled and unworthy motives attributed to them.

There is no such thing as technical infallibility, as technology too is a product of the human mind and is applied to projects by human beings. So prudence is wise and technological arrogance must be resisted. The disaster that overtook NASA's Challenger space shuttle, the Chernobyl nuclear plant leak in the former Soviet Union, and the Bhopal gas tragedy are often cited. From this follows the argument that nobody's "we-know-best" should be taken for granted and that clearances should be sought not from "interested parties" but from independent authorities. This has been sought to be done in respect of the Tehri dam. Will anyone guarantee that the Dam is absolutely safe? Probably not. No more than anyone can guarantee that no one will be knocked over by a bus or slip and fall. The test is and has to be reasonable, even abundant, caution in reducing hazards and risks to socially and politically acceptable limits because there is a cost in doing without. Every society takes such risks every day. And that is the story of civilisation. Without that mankind could not have progressed.

REFERENCES

1. Asthana, B.N. and Jain, K.K., Tehri Hydro Development Corporation Ltd. Paper presented at symposium on Large Dams *vs* Small Dams: Socio-Environmental and Techno-Economic Assessment. New Delhi, December 1991.

2. Working Group to Assess the Environmental Impact of the Tehri Dam, Report on, Ministry of Environment and Forests, New Delhi, August 1986.

3. Naqvi, Shafeeq, Seminar on Tehri High Dam, Issue Paper. Himalaya Seva Sangh, New Delhi, March 8, 1981.

4. Lal Bahadur Shastri National Academy of Administration, *Tehri Dam Project: A Report* (by Probationers). Mussoorie, October 1985.

5. Tehri Bandh Virodhi Sangharsh Samiti and Others *vs* Respondents: The State of U.P. and Others. In the Supreme Court. Writ Petition No. 12829 of 1985. New Delhi.

6. Jayal, N.D. and Chhaya, Divyang K., Written submissions on behalf of INTACH in the Supreme Court in Writ Petition No. 12829 of 1985. New Delhi, September 1986.

7. Supreme Court Order, Justice K.N. Singh and Kuldip Singh in Writ Petition No. 12829 of 1985 in TVBSS and Others *vs* the State of U.P. and Others. New Delhi, November 7, 1990.

8. Jayal, N.D. and Singh, Shekhar, Writ Petition No. .... of 1991, Supreme Court of India, New Delhi, December 12, 1991.

9. Indo-Soviet Expert Negotiations on the Issue of Cooperation in Designing and Construction of the Rockfill Dam and Spillway of the Tehri Hydro Power Complex. Protocol No. 3, New Delhi, December 25, 1988.

10. Khanna, M.M.L, Tehri Dam. Letter in *The Times of India*, New Delhi, January 22, 1988.

11. Paranjpye, Vijay, Evaluating the Tehri Dam: An Extended Cost-Benefit Appraisal. INTACH Series No. I in *Studies in Ecology and Sustainable Development*. New Delhi, June 1988.

12. Rai, Usha, USSR Not Keen on Tehri Project. *The Times of India*, New Delhi, November 15, 1988.

13. Environment and Forests Ministry, Environmental Clearance - Tehri Dam Project, Uttar Pradesh. Letter to Tehri Hydro Development Corporation. New Delhi, July 19, 1990.

14. *Bhagirath ki Pukar*, INTACH, New Delhi, February 1993 and August 1993.

15. Bahuguna, Sunderlal, An Environmentalist's Open Letter. *The Tribune*, Chandigarh, June 17, 1992.

16. Gaur, Vinod K., *Earthquake Risk to Tehri Dam*. National Geophysi-

cal Research Institute, Hyderabad, October 1984.

17. Brune, James N., *The Seismic Hazard at Tehri Dam*. Paper prepared for a discussion meeting at the Indian National Science Academy, New Delhi, December 1990.

18. INTACH: The Tehri Dam, A Prescription for Disaster. INTACH *Environmental Series No.* 6. New Delhi, 1987.

19. Thatte, C.D., *Earthquakes, Dam Design and Tehri Project*. Indian Geotechnical Society's 14th Annual IGS Lecture. Surat, December 1991.

20. Krishna, Jai, *Tall Dams in Seismic Environment*. Paper presented at Symposium on Large Dams *vs* Small Dams, Socio-Environmental and Techno-Economic Assessment. New Delhi, December 1991.

21. Chandrasekaran, A.R. and Das, Jasodhir, Analysis of Strong Motion Accelograms of Uttarkashi Earthquake of October 20, 1991. Department of Earthquake Engineering, University of Roorkee, Roorkee, December 1991.

22. Tehri Hydro Development Corporation. *Safety Aspects*. THDC Ltd., New Delhi, 1992.

23. Bolt, Bruce A., The Estimation of Seismic Risk for Large Structures in Regions Like the Himalaya. Department of Geology and Geophysics and Department of Civil Engineering, University of California at Berkeley. Science in Public Policy. Keynote Paper presented at the Workshop on Earthquake Hazard and Large Dams in the Himalaya. INTACH. New Delhi, January 15-16, 1993.

24. Iyengar, R.N., *Dynamic Analysis and Seismic Specification of Tehri Dam*. Department of Civil Engineering, Indian Institute of Science, Bangalore. Science in Public Policy: Keynote paper presented at the Workshop on Earthquake Hazard and Large Dams in the Himalaya. INTACH, New Delhi, January 15-16, 1993.

25. Reiter, Peter, Dambreak Flood Hazard Analyse (DBHA), Lecture at First Conference on Research Needs in Dam Safety, Central Board of Irrigation and Power, New Delhi, December 3-6, 1991.

26. Nielsen, N.M., BC Hydro's Approach to Dam Safety. International Water and Power and Dam Construction, Sutton, Surrey, U.K. March 1993.

27. THDC note : Some Main Features of the Project. New Delhi, 1993.

28. Administrative Staff College, *Socio-Economic Evaluation of THDC Resettlement and Rehabilitation*. Hyderabad, December 1992.

29. Dhawan, B.D., Benefit-Cost of Tehri Dam Project: A Review Analysis, *Economic and Political Weekly*. Bombay, August 31, 1991.

30. Shiva, Vandana, Conscience-Keepers of the Nation are no Terrorists. *Sunday Observer*, New Delhi, April 12, 1992.

31. Nandy, Pritish, The Old Man and the River: On Sunderlal Bahuguna. *Sunday Observer*, New Delhi, April 12, 1992.

32. Agarwal, Anil, and Narain, Sunita, Chipko People Driven to "Jungle Kato" Stir: Green Politics. *Economic Times*, New Delhi, March 31, 1991.

33. Tehri Hydro Development Corporation 4th Annual Report, 1991-92. New Delhi, September 30, 1992.

34. G.B. Pant Institute of Himalayan Environment & Development. Action Plan for Himalaya. *Himalvikas Occasional*, Publication No. 2, Almora, 1992.

35. Central Water Commission. Report on Dam Safety Procedures. New Delhi, July 1986.

5

Narmada Saga

A rock shelter painting from Bhimbetka, sculpted on the crest of the Vindhyan heights between Bhopal and Hoshangabad, depicts what art historians describe as a dyke in the Narmada Valley, one of the oldest sites of human habitation in India. The Narmada, however, appears to lack the tradition of ancient irrigation systems associated with several Indian rivers. Hemmed in by the Vindhya and Satpuda ranges, it runs through a deep valley over considerable distances in its 1312 km journey to the sea. Both topography and soils inhibited large irrigation works. Hence, despite being among the country's largest rivers, it remains among the least developed.

Investigations into harnessing the waters of the Narmada Valley commenced around the time of Independence. The integration of princely states reduced the political complexity of the task but with the subsequent reorganisation of States inter-state disputes between Madhya Pradesh, Maharashtra and Gujarat precluded progress. Work had commenced on some of the lesser components of Madhya Pradesh's Narmada master plan but the decks were not cleared for the two principal dams, the Narmada Sagar (NSP) in MP, and the Sardar Sarovar (SSP), the terminal dam in Gujarat, until the final award of the Narmada Water Disputes Tribunal (NWDT) in 1979.

The Award allocated the waters of the rivers and related costs and benefits between the riparian states, with Rajasthan being admitted as a beneficiary by agreement. Rehabilitation of those liable to be affected by the projects was addressed by the Tribunal which enunciated a package more liberal than anything preceding it, with land as its mainstay. The assumptions of land availability were soon to be over-turned by enactment of the

Forest Conservation Act, 1980, which barred any diversion of forest lands. The ensuing problems and uncertainties were followed by demands for environmental studies with a sweep and detail greater than anything required before. Even as the State governments and project authorities turned to these tasks, controversy arose and opposition grew. Overnight, the Narmada Valley projects, and Sardar Sarovar in particular, were transformed into an international *cause celebre*, a hate word. Protests, marches, dharnas, fasts, petitions and national and international pressures mounted to stop the SSP, soon to be labelled a planned environmental disaster, an engine of tribal genocide, a human rights crime and a gigantic hoax. While this reflected the strong views among activists and critics outside Gujarat, the project authorities, the Government of Gujarat and the broad mass of Gujarati opinion rose to defend Sardar Sarovar as the State's lifeline. Emotions were aroused. Eminent persons joined the debate seeking withdrawal, revision and alternatives.

Numerous internal and external conditionalities were imposed. Japan suspended aid for generation equipment. The World Bank President, Babar Conable, took the unprecedented step of instituting an independent review which, in damning the Sardar Sarovar project as virtually "impossible", stimulated fresh debate and much acrimony with charges of bias being levelled against the Independent Review Mission (IRM). The Government of India finally decided to terminate further drawals from a remaining $ 180 m World Bank credit and determined to proceed with the project through national resource mobilisation. The activists, not to be outdone, pressed for a further review and a work stoppage, otherwise threatening *jal samarpan* or death by drowning by select volunteers in the Narmada's flood waters. A "review" of uncertain dimensions was initiated by Government of India in July 1993. Gujarat boycotted the proceedings and expressed its determination to carry on with the project come what may. It held that while improvements could always be effected in the course of implementation, the basic parameters of the SSP as settled by the NWDT were legally binding for 45 years and beyond review.

This is where matters rest. And it is the course of events and the substance and rationale behind the opposing stands that we shall proceed to examine.

Early Background

It was on the eve of Independence in 1946 that the Government of the then Central Provinces and Berar requested the Central Waterways, Irrigation and Navigation Commission to investigate the potential of the Narmada river for multi-purpose basin-wide development. Several storage sites were identified and seven projects, including the Bharuch weir scheme in Gujarat, were recommended for detailed investigation. In 1948 the Government appointed the Khosla Committee to prioritise this proposal which it did, naming the Tawa, Bargi, Punasa and Bharuch projects (the last three on the main stem) for preparation of project reports. These were ready by 1963. Meanwhile, a parallel study of the hydro-electric potential of the Narmada system identified 16 sites with a potential of 1300 MW. The Central Water and Power Commission, as renamed, completed a project report on the Bharuch weir project in 1956. The Khosla Committee selected Gora as the site for a weir with a pond level of 160 feet in the first instance to irrigate 0.44 m ha in what was then Bombay Presidency. Shortly thereafter, a more favourable site about 3 km upstream of Gora was identified at Navagam village and the project reformulated to raise the dam in a second stage from a Full Reservoir Level (FRL) of 160 feet to FRL 300, with a high level canal to command additional areas. In 1959, however, a panel of consultants proposed raising the dam to FRL 320 so as to be able to carry water towards the Rann of Kachchh.

Bombay was bifurcated in 1960. The establishment of two successor States of Gujarat and Maharashtra changed the picture. A higher dam at FRL 320 at Navagam was found to submerge a potential hydel site in Maharashtra. Navagam being a terminal dam and the only site available in Gujarat, the new Gujarat Government took the view that a higher dam at FRL 460 to maximise storage would be desirable though this would submerge further hydel sites in Maharashtra. The Union Government convened a meeting in Bhopal with the Chief Ministers of Madhya Pradesh and Gujarat in November 1963 to resolve the outstanding issues. The outcome was the Bhopal Agreement which provided for a Navagam dam at FRL 425 in Gujarat; a Punasa dam in M.P. at FRL 850 with cost and power benefit sharing between M.P., Gujarat and Maharashtra; and implementation of the Bargi dam by M.P. with some assistance from Maharashtra and Gujarat (presumably towards downstream benefits). M.P. however declined to ratify the agreement, contending that the Navagam dam should not be raised above FRL 162, this being the river bed level at its border.

To break the impasse, GOI set up a High Level Committee under Dr A.N. Khosla in September 1964 to prepare a master plan for the Narmada valley, determine its phasing and recommend alternatives to the Navagam dam and optimum reservoir levels in each case. The Committee reported a year later. The master plan it produced envisaged 12 projects in M.P. and the Navagam dam in Gujarat. The optimum FRL of the terminal dam at Navagam worked out to FRL 500 with the Full Supply Level (FSL) of the Navagam canal at FRL 300. A river-bed power house (1000 MW) and a canal-head power station (250 MW) were included. The Khosla Committee gave M.P and Gujarat less water and irrigation benefits than they had sought, providing for 2.63 m ha irrigation in MP, 40,000 ha in Maharashtra, 1.85 m ha in Gujarat and 400,000 ha in Rajasthan. It gave irrigation priority over power though providing for generation of 951 MW of hydel energy. The higher Navagam dam it favoured was seen to ensure additional storage, greater carry-over capacity, increased power production and optimal irrigation and flood control. It argued the case for a high level canal to obviate lifting water and carrying irrigation supplies up to the international border with Pakistan. The storages it recommended in M.P. included Bargi, Tawa and Narmadasagar (Punasa) while its proposed Navagam high dam submerged the hydel sites of Jalsindhi (Maharashtra) and Harinphal (M.P.) but without any more submergence than would be caused by the three dams if separately constructed.

Gujarat endorsed the Khosla Committee's findings. M.P. and Maharashtra rejected them. M.P. opposed the high dam and the reduced water allocation and loss of hydel sites in favour of Gujarat. It was for sharing the benefits of the Jalsindhi hydel project with Maharashtra and said that it was prepared to make certain regulated releases of water to Gujarat to meet its irrigation needs on a cost-sharing basis provided the Navagam dam was lowered. It made out a case for a larger water allocation on the plea that it had identified attractive trans-basin diversions within its own boundaries. Official and ministerial parleys having proved unavailing, GOI decided to refer the Narmada dispute to a Tribunal under the Inter-State Water Disputes Act.

Basin Description

The Narmada rises at Amarkantak in Shahdol district in eastern M.P. at an elevation of 900 m. It is held sacred and dotted by famous shrines and is the scene of the holy Parikrama pilgrimage. The river cascades over several falls especially in its upper reaches and flows through a number of gorges, including the renowned Marble Rocks near Jabalpur. It is joined by 41 tributaries, over its length of 1312 km, 22 from the Satpuda and the rest, on the right bank, from the Vindhya range. Of the 98,796 sq. km. basin, roughly 86 per cent is in M.P., 2 per cent in Maharashtra and the rest in Gujarat. Precipitation ranges from 1200 to 1600 mm in the upper catchment to half that figure in the lower region, the lushness and diversity of vegetation declining from east to west. The Irrigation Commission (1972) identified parts of M.P. within the Narmada basin as being drought affected. It also defined a large part of North Gujarat, Saurashtra and Kachchh as semi-arid or arid scarcity regions on account of the "extreme unreliability" of rainfall, rendering them "chronically drought prone" and subject to serious drinking water problems. The Commisssion noted that "Gujarat has suffered famine and scarcity in 23 of the 71 years of the present century. The countrywide drought of 1965-66 and 1966-67 affected 5–6000 villages in the State .... In the 1968-69 drought, 8.8 m people in 10,000 villages were affected; in 6000 villages the situation was acute".

A good part of the Narmada basin has light, medium and heavy vertisols (black cotton soils). Of the 4.5 m ha net area sown in the basin only a little over 200,000 ha was irrigated in 1972.

With the reference of the Narmada water dispute to a Tribunal, Madhya Pradesh in 1970 put together an expert group to prepare a Master Plan for the development of Narmada waters in the State. This reported in 1972 and suggested the construction of 31 major dams on the river and its tributaries above Navagam. It was hoped that 10 of these would be completed within a decade and would moderate floods. Navigation too was envisaged with the harnessing of the river.

The Narmada dispute was closely contested before the Tribunal for a whole decade, 1969-79. Earlier plans were held in abeyance, overtaken by events or abandoned. In the last category was the Bharuch Weir Scheme (FRL 160) and low level canal for which Jawaharlal Nehru laid the foundation stone at Gora, just below the present SSP dam site, in 1961. Second thoughts on the height of the dam resulted in a pause and subsequently a change in site. It was as part of this project that the Kevadia colony, now the base camp for the SSP, was established and the first displacement occurred. The SSP rehabilitation story starts from there.

Issues before the Tribunal

Each of the concerned States raised a number of issues before the Tribunal, many of them conflicting, with regard to riparian rights, heights of proposed dams that might prejudice or submerge other dam sites, and even whether there was a dispute at all in terms of the River Water Disputes Act. Gujarat pleaded that Maharashtra and M.P. had concluded agreements to construct certain dams which would limit the height of the terminal dam it proposed to build to FRL 210, thereby diminishing the irrigation and power benefits and precluding reclamation of arid tracts in the Ranns of Kachchh. M.P. demurred. It argued that the planned hydel projects entailed no consumptive use of water to Gujarat's detriment. On the other hand, Gujarat had no right to raise the height of the Navagam dam and submerge territories beyond its borders, including the Jalsindhi, Harinphal and Maheshwar sites. Nor was Rajasthan a co-riparian state and could not therefore make any claim to Narmada waters.

The Tribunal framed 24 issues for preliminary consideration. These included the quantum of utilisable Narmada flows at Navagam; principles of equitable apportionment between the parties concerned; whether preference be given to irrigation over power; should diversion of water outside the Narmada basin be permitted and, if so to what extent; did the States concerned have alternative means of satisfying their needs; were there any existing appropriations of water and, if so, to what extent should they be protected; how should costs and benefits be shared; if the

Navagam dam were to submerge areas in Maharashtra and M.P. what directions should be given to them to acquire land for the purpose and on what terms of compensation, including rehabilitation of displaced persons; and, finally, what machinery should be set up to oversee implementation of the Tribunal's order.

The Tribunal delivered its judgement on the preliminary issues in February 1972. It held that there was a water dispute, and that it would be in order to consider Gujarat's plaint for a high dam on the basis of compensation for any injury caused. It ruled that Rajasthan had no *locus standi* and its complaint was *ultra vires*. The Governments of MP and Rajasthan appealed to the Supreme Court and obtained a limited stay on the proceedings of the Tribunal. However, in July 1972, the Tribunal was informed that all the four States had entered into a compromise agreement through the good offices of the Prime Minister. They accordingly sought an adjournment of further proceedings. The resulting discussions led to an agreement in July 1974. This stipulated that the quantum of water available for allocation be assessed at 28 million acre feet (MAF), that the requirements of Maharashtra and Rajasthan be taken as 0.25 and 0.5 MAF respectively and that the balance of 27.25 MAF be shared between M.P. and Gujarat on the basis of a determination by the Tribunal of the height of the Navagam dam. The parties thereupon withdrew their petition from the Supreme Court and the Tribunal decided to accept the mutually agreed settlement.

Since the Tribunal was required to allocate the waters of the Narmada between the contending States, its first task was to determine the quantum of water available. This figure is normally obtained by ascertaining the mean of a long flow series that evens out annual and cyclical fluctuations in rainfall and runoff. Further, for planning purposes, the Indian norm has been to take a figure of 75 per cent dependability or that flow which can be depended upon to occur or be exceeded in 75 out of every 100 years, shortages in the remaining 25 years being made good to the extent possible by any carry-over from more generous flows during previous years.

The problem faced by the Khosla Committee when it started work on preparation of a Master Plan for the Narmada basin was the limited availability of discharge readings from 1948 to 1962.

This gave a flow at Garudeshwar (12 km below Navagam) of only 22.60 MAF. Being too short a series to establish dependability, the Committee resorted to the standard international practice of hindcasting on the basis of rainfall data which was fortunately available from a number of stations since 1915 with given rainfall-runoff relationships established for different zones. The runoff thus derived was 28.92 MAF. Maharashtra and M.P. independently worked out a runoff hindcast to 1891 at 27.14 and 27.17 MAF respectively. Controversy was stilled with the agreement by the four Chief Ministers in July 1974 on a technically assessed runoff of 28 MAF. The Tribunal accordingly adopted this figure.

The Tribunal saw no reason to include the area already irrigated by the Mahi project within the Narmada command. It also excluded the Ranns and the Banni area of Kachchh in determining an equitable apportionment of water betweeen M.P. and Gujarat. However, it set aside 1.52 MAF and 1.06 MAF of water for domestic and industrial requirements up to 2021 in M.P. and Gujarat respectively. M.P. plans to supply Bhopal, Indore and Mhow with drinking water from the Narmada, while Gujarat seeks to ensure domestic supplies to all of North Gujarat, Saurashtra and Kachchh from this source.

Gujarat had claimed 1000 cusecs year-round or 0.7 MAF annually for releases below Navagam for navigation, domestic water requirements and irrigation. This was disallowed. But the Tribunal took account of 0.282 MAF of water likely to be available to Gujarat from rivers crossing the Navagam Canal en route. Noting that power could be generated through other means, the Tribunal gave priority to irrigation over hydel generation. On an examination of various heights and combinations of dams, including Navagam, Jalsindhi and Harinphal, the water requirements of Gujarat and the FSL of the Navagam Canal if it was to serve the approved command, the Tribunal determined the height of the Navagam dam despite submergence of the Jalsindhi and Harinphal sites.

The decision of the Tribunal was announced in August 1978. Within three months of that the Union of India and the four State Governments concerned filed five separate references seeking clarifications. These were considered by the Tribunal which issued its final orders in December 1979.

NWDT's Final Order

The Tribunal's final order determined the utilisable quantum of Narmada waters to be 28 MAF at 75 per cent dependability. Of this, M.P. was allotted 18.25 MAF, Gujarat 9 MAF, Maharashtra 0.25 MAF and Rajasthan 0.5 MAF. It was made clear that each State will be free to make such changes in the pattern of water use and in the areas to be benefited, within or outside the Narmada basin in its own territory, as it deems fit. Utilisable flows in excess of 28 MAF, including carry-over flows, will be shared in the ratios of 73 to MP, 36 for Gujarat, 2 for Rajasthan and 1 for Maharashtra. Shortages shall be shared in the same proportions. The utilisable flows are to exclude evaporation losses at the reservoirs but will include all return flows. If any reservoir spills, the concerned States shall be entitled to utilise the surplus in any manner without this establishing any prescriptive rights.

The apportionment of water ordered is to be reviewed after 45 years. The Navagam Dam height will have a FRL of 455 feet with a Maximum Water Level of 460 feet. The FSL (full supply level) of the Navagam Canal was fixed at 300 feet with prescribed gradients. With regard to the sharing of costs and benefits, the Tribunal determined that the net power produced by the (Sardar Sarovar) Navagam Dam and canal shall be shared in the ratio of 57 per cent to M.P., 27 per cent to Maharashtra and 16 per cent to Gujarat and that the cost of construction allocable to power shall be shared in the same proportions and paid in annual instalments. M.P. in turn is required to make regulated releases from the Narmada Sagar Dam to meet Gujarat's allotted supplies broken down into 10-day periods. For this downstream benefit, Gujarat shall pay M.P. 17.63 per cent of the cost of the Narmada Sagar Dam in annual instalments from the commencement of construction.

The Resettlement and Rehabilitation (R & R) scheme set out by the Tribunal in relation to land acquisition and displacement was among the most liberal devised at the time. The Tribunal defined an "oustee" as any person who one year prior to the publication of a notification under Section 4 of the Land Acquisition Act has been ordinarily residing, farming or working in the submergence area up to FRL 455. Buildings shall, however, be acquired in Maharashtra and M.P. for the SSP up to MWL 460, including appurtenant land, subject to backwater effects. The maximum backwater level shall be determined by the Central Water Commission. Gujarat shall pay Maharashtra and M.P. compensation for all government lands acquired, the costs of rehabilitation of oustees within their own territories and the expenses incurred on the related establishment required to execute this task, as well as the costs of relocating any ancient or historic monument, places of worship or idols likely to suffer submergence.

The Tribunal stipulated that every major son shall be treated as a separate family and on that basis make a rough count of likely oustees. It held that displaced persons in Maharashtra and M.P. desirous of resettlement in Gujarat must be rehabilitated there. Oustees were entitled to a 60 × 90 feet homestead plot, a prescribed scale of grants and civic amenities (such as a school, panchayat ghar, dispensary, seed store, children's park, village pond, drinking wells, road links and place of worship). Every displaced family from whom more than 25 per cent of his holding is acquired shall be entitled to a minimum of two ha of irrigable land, the price to be paid in 20 interest-free instalments, 50 per cent of the compensation received being set off as the initial deposit. It was provided that in no event shall any area in Maharashtra and M.P. be submerged until compensation is paid and arrangements for R & R completed.

Gujarat was required to furnish Maharashtra and M.P. a schedule of submergence linked to construction of the dam 18 months in advance. The residents, would, however, not be required to vacate the area until six months before submergence but "must vacate the area by the notified date". While the Award is binding, the Tribunal held that amendments, alterations and modifications may be affected "by agreement between all the party States". A Narmada Control Authority was to be established to implement the decisions of the Tribunal and the Award was made subject to review at any time after 45 years. (NWDT, 1979)

Scope of Narmada Valley Development

While the current controversy about the "Narmada project" centres primarily on the terminal Sardar Sarovar Dam in Gujarat and, to a lesser degree (as yet), on the interlinked Narmada Sagar (now Indira Sagar) Dam in Madhya Pradesh, these have to be seen in the context of the overall plan for development of the water resources of the Narmada Valley, home to over 40 million people, a goodly proportion of them tribal.

Ultimately conceived are 29 major dams, 135 medium dams and over 3000 minor dams in Madhya Pradesh to utilise the State's share of 18.25 MAF of water to irrigate 2.75 m ha of land, generate 2600 MW of power and provide municipal, industrial and cooling water in the Valley. Of these projects, the Tawa, Barna and Sukta major dams are complete while the Bargi, Mateyari and Kolar projects are at an advanced stage of construction. The Man and Jobat projects are at an initial stage. The kingpin of the Madhya Pradesh master plan, however, is the Narmada Sagar Complex consisting of the principal Indira Sagar Dam near Punasa, and below it, in a cascade, the Omkareshwar multipurpose project (283,000 ha/520 MW) also in Khandwa district, and the Maheshwar hydel project (400 MW) in Khargone district. The three dams will between them store 11.2 MAF of water (gross), 9.9 MAF of this at Indira Sagar which will irrigate 265,000 ha and have an installed capacity of 1000 MW. Firm power in all cases will be less and will decline as irrigation picks up. Omkareshwar and Maheshwar will utilise regulated releases from Indira Sagar which will guarantee 8.12 MAF to Sardar Sarovar below Maheshwar. The gross submergence area for the Narmada complex as a whole is estimated at 105,607 ha (91,348 ha under Indira Sagar and variously affecting 249 villages), displacing a population of around 100,000 persons.

The Sardar Sarovar Project (SSP) is far more ambitious. Its proposed benefits include irrigation of just under 1.8 m ha in Gujarat and 75,000 ha in Barmer district of Rajasthan through a 425 mile long Main Canal with a capacity of 40,000 cusecs at the offtake, making it the biggest in the world. Additionally, it will have an installed hydel capacity of 1450 MW and will provide drinking water to 131 urban centres and 4720 villages, including all of Saurashtra and Kachchh, drought proof a large part of Gujarat, provide flood protection to a population of 750,000 in 210 villages and to Bharuch city, and stimulate afforestation, tree plantation and aquaculture. As against this, it will submerge 34,867 ha of land, 7469 ha in Gujarat, 6570 ha in Maharashtra and

20,828 ha in M.P. Over a third of the total area submerged will be river bed and waste land, 11,279 ha agricultural land and 10,719 ha forest land. Some 248 villages will be submerged, 245 of them partially, affecting a total population of around 100,000.

Critics would have it that the benefits have been exaggerated and the costs concealed and that the ultimate outcome will be a human, ecological and financial disaster. Official sources claim that seldom has a project been so long and so thoroughly studied and investigated at every turn as the Narmada Valley projects, and the SSP in particular. Alternatives have been examined. Much has been learnt from past experience so that both at the technological and R & R level the SSP represents a great advance. What has emerged therefore is an optimised programme with a continuing thrust towards upgradation and innovation even in the course of implementation. The other view is that there has been little or no consultation, a lack of transparency, callousness, technological conceit and window-dressing. It is to an examination of these rival contentions that we now turn.

Environmental Conditions

The NWDT enabled Gujarat and M.P. to go ahead with detailed project designing and related studies on the basis of the stipulated parameters. The World Bank was drawn in soon thereafter and recommended the formation of an inter-disciplinary Narmada Planning Group (NPG) as well as a number of independent studies for the SSP. The Government of Gujarat constituted the NPG which in turn initiated and farmed out a number of studies involving several major institutions in the State and established consultants. Preliminary work was started. The Tata Economic Consultancy Services turned in an economic appraisal which assessed the benefit-cost ratio of SSP at 1.84 at economic prices (1983) a figure which the NPG subsequently estimated at 1.94 with a 12 per cent discount rate (1989).

Environmental impact and socio-economic studies having come in from both Gujarat and M.P. the Central Water Commission approved the projects in 1984. Preliminary works commenced. The World Bank undertook Staff Appraisals in 1985 and 1986 as a prelude to concluding loan agreements for the SSP and NSP. With regard to the SSP, the Bank noted that within a period of three years an impressive array of high quality studies and designs had resulted, "including a comprehensive framework plan by the NPG and production by the ID (Irrigation Department) of high quality designs, specifications and procurement documents" all of which "represents a break with past approaches to the planning, design, construction and operation of irrigation projects in India". (NPG, November 1989)

The World Bank found the NWDT's R & R package to be an advance on previous Indian norms but nevertheless falling short of its own evolving standards. The $ 450 m loan agreement it entered into with the Government of India (GOI) on behalf of the Governments of Gujarat (GOG), Maharashtra (GOM) and Madhya Pradesh (GOMP) included landless agricultural labourers and encroachers in the category eligible for 2 ha compensation in respect of all SSP project-affected persons (PAPS) in the three States. It also made loan disbursements conditional on time-bound completion of more detailed environmental impact assessments and remedial plans. (Paranjpye, June 1990) Consideration of assistance for the NSP was deferred, though simultaneous construction of this with the SSP had been envisaged as both projects are expected to work in tandem. (IRM, June 1992)

Two significant developments had meanwhile taken place after the NWDT passed its final order. The submissions made before the Tribunal assumed that degraded forest lands would be made available by Government for rehabilitation. However, the passage of the Forest Conservation Act in 1980 precluded any such possibility as diversion of forest lands (let alone forests) to non-forest use was expressly barred except through special dispensation of the Government of India. At a broader level, a Ministry of Environment (and Forests) had come to be established as the newly-found conscience and custodian of the country's environmental interests. Inexperience and zeal combined to make a heady brew and the new Department acquired considerable clout from its ability to say "no". Environment impact studies gradually came to be required on the basis of evolving guidelines and environmental clearances became mandatory in the procedural progression of projects to investment sanction. Environmental conditionalities might be imposed if considered necessary.

In April 1986, the Ministry of Environment and Forests (MOEF) came up with a paper on the Environmental Aspects of the Narmada Sagar and Sardar Sarovar Multi-purpose Projects. This argued that it would take another two to three years to complete environmental impact studies for these two projects; that the environmental impacts of the NSP were likely to be more severe than those of the SSP; only a third of the lands required for R & R had been identified and their land capability remained unsurveyed; and that considerable submergence and displacement could be avoided by a modest reduction in the height of the NSP without affecting irrigation from the SSP if the minimum draw-down level of the latter were correspondingly reduced. (Paranjpye, June 1990)

Environmental critics took the cue and protests came to be voiced. The Water Resources Ministry (MOWR), GOG and GOMP, however, pressed for environmental clearance for both projects so as not to impede the momentum built up thus far, avoid delays and ensure simultaneity in construction. MOEF was prevailed upon to relent and finally did grant conditional environmental clearance on June 24, 1987 for both NSP and SSP and forest clearance (for the SSP alone) on September 8, 1987.

The Memorandum on Environment Approval noted that the Environmental Appraisal Committee for River Valley Projects had sought details of both projects with regard to their rehabilitation master plans, phased catchment area treatment (CAT), compensatory afforestation, command area development (CAD), flora and fauna surveys, the carrying capacity of the surrounding areas, seismicity, and health aspects. It noted that field surveys were yet to be completed and that full details must be furnished by 1989. The Narmada Control Authority (NCA) was required to ensure that environmental safeguards were planned and executed *pari passu* with implementation of the engineering works. The NSP and SSP were accorded environmental clearance on the further condition that compensatory afforestation and R & R programmes would be completed before reservoir filling and MOEF supplied with periodic information reports. A separate forest clearance was to issue.

Forest clearance was issued 10 weeks later in response to

applications made by GOG and GOM in 1983 and GOMP in 1984. Diversion of 13,385.45 ha of forest land was approved for diversion or submergence under the SSP, 4165.91 ha of this in Gujarat, 2,731 ha in M.P. and 6,488.54 ha in Maharashtra. The conditions stipulated were that the legal status of the land would remain unchanged; full details of the non-forest lands earmarked for compensatory afforestation would be provided before September 30, 1987 and these areas handed over to the respective State Forest Departments and declared reserved/protected forests; non-forest areas available for R & R or "a proposal to the satisfaction of GOI" would be communicated by November 30, 1987; no work in forest areas was to commence before these conditions were fulfilled; "since the project involves violation" and most non-forest areas for compensatory afforestation are away from the project area, the State Governments concerned must undertake compensatory afforestation in double the degraded forest lands lost and also in the project impact areas in addition to afforestation on equivalent non-forest land, with a scheme to this end to be submitted by November 30, 1987; a CAT plan was to be prepared by March 31, 1989 and implemented at the cost of the project; no forest land would be utilised for R & R; tree felling in the reservoir would be permitted only up to four metres below FRL; tree planting would be done along canals, roads, the reservoir foreshore and on waste/vacant lands with the State Irrigation Department; fuelwood depots must be set up by the project authorities and alternative fuels supplied free to all project labour and at subsidised rates to other staff; and, finally, a wildlife conservation and management committee was to be constituted by the State Governments by January 31, 1989 to draw up a suitable plan to be implemented at project cost.

The deadlines given were short; the burdens onerous.

The Planning Commission's approval of SSP, given on October 5, 1988 mentioned the need for compliance with the conditionalities accompanying the environmental and forest clearance. The SSP was sanctioned with an investment cost of Rs. 6406 crores. GOG was required to give it priority funding during the Eighth Plan and to complete other on-going projects that were at an advanced stage of construction so as to ensure funding of the SSP's peak requirements. A programme of drainage and groundwater balance studies was to be completed for command areas beyond the Mahi and submitted for monitoring by the Commission. Suitable water rates were to be devised to ensure annual revenues from the project sufficient to cover annual operating costs. Canal siltation studies were to be undertaken. And finally, micro-level canal network plans and estimates were to be got ready within five years and the entire system planned to be completed in segments up to farm outlets so as to ensure early utilisation of available irrigation supplies. In according investment sanction the Planning Commission also, therefore, laid down a number of environmental conditionalities.

Evolving Rehabilitation Norms

The NWDT, as already mentioned, laid out a rehabilitation package which, certainly as far as landholders were concerned, was in advance of previous practice. This, however, did not cover other land-related categories such as agricultural labourers, encroachers and co-parceners. Nor did it make provision for canal-affected persons. The World Bank moved to fill these gaps incrementally as its own policies evolved in the light of international experience. The NWDT package, however, only applied to those affected by SSP submergence provided they opted to settle in Gujarat. This, therefore, did not bind the M.P. and Maharashtra governments to adopt identical policies for those of their residents choosing to resettle locally rather than move to Gujarat. Further, with the enforcement of the Forest Conservation Act, diverting "forest lands" (an administrative term signifying departmental jurisdiction rather than actually forested areas) for R & R was prohibited, rendering many potential sites out of reach. To state this is not to extenuate any failure but to underline changing ground rules and draw attention to the complexities involved.

Cash compensation, as required under the Land Acquisition Act, has been found to be unsuitable among simple peasant or tribal communities who often live in a semi-monetised economy and are unable to handle large sums of money. Hence the advocacy of land to agriculturists and all others in any way dependent on the land, especially tribals. This is understandable though not always practical, with growing population pressures and a falling man-land ratio. Host populations among whom those displaced are resettled are often likely to possess less land than the new settlers — two hectares apiece as a minimum in terms of the NWDT's package for Gujarat. Fortunately, Gujarat's farmers are not without entrepreneurial aspirations and have been found to be willing to sell their lands and use the capital to invest in various avenues of self-employment. The existence of such a socially and economicaly mobile rural population has engendered a land market which the Sardar Sarovar Nahar Nigam Ltd. (SSNNL) has been able to tap through land purchase committees and make sites available for R & R. These lands have in fact been inspected and selected by the Project affected persons (PAPs). The situation elsewhere may not necessarily or always be quite so encouraging.

It would not therefore be desirable to insist on land for land, though land is an obvious means of rehabilitation for rural/ peasant communities if and where available. The more practical and, ultimately, productive alternative is employment backed by education and training, especially in view of the lead time available in water resource schemes and the temporary and permanent work opportunities, direct and indirect, they provide. (Verghese, 1990) However, a strong commitment to land for land has been made in the case of the Narmada Valley Projects and it may neither be possible nor prudent to backtrack at this point of time. Whatever commitments have been made should therefore be fully honoured in letter and spirit. The NWDT prescribed a minimum 2 ha (five acre) holding for SSP land oustees and this has been sought to be made the norm by the World Bank and others for extended land-related categories such as major sons, landless agricultural labourers, encroachers and co-parceners. This has placed an onerous burden on the Project authorities and State governments to find land in non-forest areas. Other ongoing World Bank aided projects, such as the Upper Krishna in Karnataka, prescribe lower land limits as compensation for landed oustees and even less for others.

The SSP has been caught on an escalator of evolving environmental-cum-land rehabilitation policies both on the part of GOI and the World Bank. Reference has already been made to the new situation created after the NWDT package was announced.

with a prohibition on diversion of forest land under the Forest Conservation Act, 1980. The World Bank in February 1980 published an operational manual statement on Social Issues Associated with Involuntary Resettlement in Bank-Financed Projects. The thrust of this was that "after a reasonable transition period, the displaced people regain at least their previous standard of living and that so far as possible, they be economically and socially integrated into the host communities". (IRM, 1992) The policy was not confined to dams but extended to "canals, highways, transmission lines and the like". This was followed in February 1982 by an operational manual on Tribal People in Bank-Financed Projects. This noted the need to safeguard (traditional) tribal interests, avoiding extreme measures either perpetuating isolation or promoting "forced accelerated acculturation unsuited to the future well-being of the affected tribe". It showed awareness of the dilemma facing governments in harmonising, say, a hydro-project with the rights of tribal oustees and said such issues were "matters of judgement guided by the principle that Bank assistance should help or mitigate harm, and provide adequate time and conditions for acculturation."

The statement is unexceptionable and accords fully with India's established constitutional and other guarantees to its Scheduled Tribes which have long preceded and in many ways remain in advance of international concerns about tribal and indigenous people which have been more recently articulated. The Bank's loan and credit agreements for SSP in 1985 gave it leverage to persuade the GOI and concerned State governments to move towards adoption of some or all these norms. The prime criterion remained that those displaced be enabled to regain or improve their living standards within a reasonable period of time and that they be offered a choice of location and be participants in the process of implementation. An important extension of principle lay in bringing in those ordinarily residing and cultivating in the submergence area within the (Bank's) meaning of land-related oustees, including landless agricultural labourers and encroachers. A 1990 operational directive of the Bank specifically provided that "absence of legal title to land by such (tribal) groups should not be a bar on compensation". Finally, in September 1991, the World Bank issued yet another operational directive regarding tribal and indigenous people whose "dignity, human rights and cultural uniqueness" were entitled to full respect. Taken together, these directives represent a decadal policy evolution regarding involuntary displacement of tribal populations and others. This recital, however, cannot obscure the fact that GOI and State governments were legally bound by the NWDT and at best by the specific terms of the 1985 World Bank loan agreements. The rest was a matter of persuasion and adjustment. Quite independently of that, however, official policy in the Narmada basin was also evolving. This was partly in response to NGO pressures to interpret the NWDT package liberally rather than restrictively and, indeed, to go beyond it. And this is what has happened.

If the SSP rehabilitation package evolved over time, it also did so in response to World Bank conditionalities which were themselves shaped in their Narmada Valley context by the insights and demands of Indian critics, notably the Action Research in Community Health Association (ARCH) Vahini under the leadership of Dr Anil Patel, and the Anand Niketan Ashram under a Gandhian, Harivallabh Parikh. Both NGOs had been working among the tribals in the vicinity of the Sardar Sarovar Dam. The Anand Niketan Ashram had been in existence for some decades and is widely known for innovating a system of lok adalats or people's courts for settlement of tribal disputes so as to avoid feuding, litigation and indebtedness. Both NGOs understandably got drawn into monitoring and then intervening to secure improvements in the SSP R & R package on behalf of their tribal constituents.

ARCH entered the fray in 1980 and over the next few years organised the 19 submergence villages in Gujarat to assert their rights under the NWDT Award and subsequent World Bank loan agreements. ARCH found the initial response of GOG falling short of NWDT norms in some respects. It led a protest march of affected tribals to the SSP Project Headquarters at Kevadia, demanding a minimum of 2 ha of land to each landed oustee, major son and encroacher. Project work on a dyke appurtenant to the dam was stopped, the Gujarat High Court was petitioned and a campaign launched to lobby the World Bank. GOG bowed to these demands and a radical new R & R package was announced by it in December 1987 which ARCH endorsed.

(ARCH, April 1991) The contribution made by ARCH and Anand Niketan is acknowledged in the World Bank's records as set out in the Report of the Independent Review Mission. (IRM, 1992)

By 1985 the movement for clarity in and improvement of the R & R package had spread to the submergence areas in Maharashtra under the Narmada Dharangast Samiti (NDS) based in Dhule and the M.P.-based Narmada Ghati Navnirman Samiti (NGNS). Their demands, set out in a 38-point charter in November 1987 to the Narmada Control Authority, Indore, focussed on R & R issues with little or no reference to environmental questions.

Another Delhi-based institution, the Multiple Action Research Group (MARG) surveyed the submergence villages in Barwani Tehsil, Khargone district, M.P. It found the level of information among the people "lamentably low" and noted poor interaction between Project authorities and the affected population. Initial project surveys had been conducted allegedly without proper notices that people knew of or understood, leaving behind stone markers of submergence levels that had aroused deep anxieties and uncertainties. MARG found some villagers ill-informed, others misinformed. They were not fully aware of their rights if they opted to go to Gujarat and even less aware of their right to be rehabilitated in M.P. itself. Six forest villages had been totally omitted from the rehabilitation lists though marked in the submergence area maps. Meanwhile, there was a feeling that even existing civic services were being withdrawn from these "doomed" villages. Also left out were landless persons, who form part of the social network through the *jajmani* (mutual service) system, female family-heads and widows. (MARG, May 1987)

Under these multiple pressures and the prodding of the World Bank, GOM and GOMP slowly began to respond and improve their R & R packages to approximate to the more liberal norms by now adopted by Gujarat. Maharashtra was handicapped, its plans having gone awry with permission to divert adjacent forest land, as planned, being now denied under the Forest Conservation Act. Over and beyond this, however, both Maharashtra and M.P. found themselves disadvantaged for another significant reason. As far as Gujarat is concerned, the SSP is about its last major project and has, not without reason, been termed the State's lifeline. It is also the major beneficiary of the SSP, through flood moderation, as the recipient of the lion's share of irrigation, and as a co-sharer in the power benefits with Maharashtra and M.P. At the same time, of the 37,533 ha submerged, as much as four-fifths of the area and as many as 226 of the 245 affected villages are in M.P. and Maharashtra.

Maharashtra and M.P. each have a considerable water resource potential waiting to be harnessed in a number of river basins. They are apprehensive of conceding in the case of SSP what could become precedents for other on-going and, certainly, future projects. In fact, since displacement is liable to occur in case of land acquisition for a variety of "public purposes" such as industry, mines, power and transportation projects, ports, urban expansion and other developments, there could well be a demand for universal application of SSP norms. This has in fact already begun to happen, with Tehri Dam oustees seeking upgradation of their R & R package to the 2 ha SSP norm.

The caution exercised by Maharashtra and M.P. can therefore be well understood. And, given states' rights in a federal polity, there are limits beyond which they can be pushed. The World Bank tacitly recognises this. The Independent Review did not and thought nothing of derogating from federal principles though both its members, Bradford Morse an American, and Thomas Berger, a Canadian, come from federal countries. This is not to say that federal principles can ride roughshod over social justice. The NWDT, a Tribunal headed by a Supreme Court judge, set out certain norms which are being observed. All three SSP-related States have improved upon the Award, Gujarat most of all and Maharashtra and M.P. in varying measure. The Independent Review's argument apropos Maharashtra's treatment of landless agriculturists and encroachers is that "a policy that offers less than Gujarat, while applying to a population with comparable socio-economic characteristics, is vulnerable to the charge of non-compliance (with the 1985 Bank loan agreements)". Hence, "if Gujarat was right to extend its policy provisions, then Maharashtra is wrong not to have followed suit". This is a non sequitur. Maharashtra and M.P., as much as Gujarat, are committed to the proposition that those involuntarily displaced should be as well if not better off as a result of R & R. It does not follow that all must take the identical route to achieve this goal.

Indeed, Gujarat itself has not treated the Kevadia Colony (1961) and so-called dyke-village displaced persons (1981) on par with those displaced after its upgraded 1988 R & R package was announced. Gujarat told the IRM that retrospective application would create "a potentially unmanageable precedent" and "open up 30 years of history". So while improvements were made to ensure justice, uniformity was not conceded.

Yet another factor was at work. Although the NWDT made it Gujarat's responsibility to resettle all displaced persons in the SSP command or elsewhere in the State, those affected were free to choose to resettle in their home states at Gujarat's expense. But then the home state would have the onus of finding the land for R & R. However, while the NWDT norms are the same in all three states, the improvements introduced by each are specific to their own territories. It was always expected by rule of thumb, and reinforced by spot verification, that tribal PAPs living in close proximity to the SSP dam in M.P. would in all probability opt to settle in Gujarat in view of cultural affinities, while those living further away in Nimad would be more inclined to be relocated nearer home within M.P. But many choices would be likely to be influenced by the relative R & R packages. The Maharashtra PAPs had all along expressed a preference for relocation in adjacent forest areas in Maharashtra, which is what they had originally been promised until the Forest Conservation Act suddenly became a hurdle.

In the circumstances, the other two States, and M.P. in particular, with much of its submergence coming later in the construction sequence, was in no hurry to announce even such improvements as it was inclined to make, in the hope that delay might cause larger numbers to opt for settlement in Gujarat. The World Bank on its part kept dangling, but did not conclude, prospective Narmada Valley environment and ISP loans, hoping thereby to encourage M.P. to hasten and improve action on its SSP commitments which would, in turn, set the pattern for the R & R package for the Narmada Sagar complex.

A comparative review of the R & R packages in the three states as they stand today indicates marked advances by each over the NWDT Award. Properly implemented they may be expected to ensure a better quality of life for those affected in a reasonable period of time. Most if not all of them will be lifted from below the poverty line, where the majority and virtually all the tribals are now trapped, to a status well above that level.

Current R & R Package

The salient features of the R & R package as evolved by the three SSP-affected States as of June 1993 (and even a year earlier when the IRM reported), are in each case an improvement on the NWDT Award. Each landed oustee will receive a minimum of 2 ha of irrigable/irrigated land of his choice either in Gujarat or in his respective home state if he so prefers, subject to a maximum of 8 ha or the State's land ceiling as the case may be. While submergence areas in excess of 25 per cent of the landholding will be acquired, any residual holding which may be economically or socially unviable will also be acquired as will elevated patches which become islands in the reservoir (tapu lands). All joint holders or co-parceners of land titles are treated as landholders.

The NWDT did not recognise encroachers on forest or non-forest lands or the landless as entitled to land benefits and initially none of the affected States was inclined to reconsider this matter. But the World Bank (and later the IRM) kept pressing the point that tribal communities have seldom held land titles and with the state declaring forests to be "reserved" or "protected" over the past century, their traditional and customary land rights have been extinguished. Moreover, owing to inaccessibility, ignorance and corruption, land transfers, inheritance and similar transactions are often unrecorded by patwaris or in settlement surveys and it would be grossly unfair to treat legitimate tribal families, cultivating lands traditionally considered theirs, as "encroachers" with no rights. Likewise, landless agricultural labourers would lose their livelihood once the lands on which they are employed were submerged. They too must therefore be provided land in recompense.

There is some merit in these arguments and Gujarat had by 1988 accepted both propositions. In all states, encroachers are periodically recognised and recorded and Gujarat said it would give land benefits to all encroachers in the submergence area a year prior to notification of the area under Section 4 of the Land Acquisition (LA) Act. Both encroachers and landless would therefore receive 2 ha of land, with encroachers being entitled to ex-gratia compensation for encroached lands in excess of this figure. Maharashtra subsequently agreed to grant 2 ha of land to encroachers and ex-gratia compensation for any excess encroached upon up to March 31, 1978. Later encroachers would be treated as landless and be given 1 ha of farmland. M.P. followed suit with a cut-off date for encroachment prior to April 13, 1987, with allotment of a hectare for those encroaching up to that limit and 2 ha for those with larger encroachments.

While Gujarat offered 2 ha to all landless, Maharashtra came round to offering one hectare free of charge if the oustee moves to the relocation site. M.P., however, decided that all landless agricultural labourers and all other landless Scheduled Caste and Scheduled Tribe oustees would be given a sum of Rs. 11,000 as a rehabilitation grant plus Rs. 29,000 to purchase any productive asset for self-employment. All other landless would be entitled to receive Rs. 5,500 and Rs. 19,500 respectively.

The NWDT provided no land benefit to major sons of landholders, encroachers or landless even while defining them as separate families for purposes of homestead plots and calculation of community benefits to be provided on relocation in accordance with the number of families in any cluster. The IRM contested this interpretation and insisted that the NWDT's directions implied treating major sons as separate families entitled to land benefits. The World Bank too had been pressing for this. In the result, Gujarat granted major sons full entitlement to land benefits subject to a cut-off date of January 1, 1987. Madhya Pradesh agreed to recognise major sons as separate families entitled only to cash compensation (not land) according to the category to which they belong. Maharashtra decided to recognise major sons and major unmarried daughters, subject to a cut-off date of January 1, 1987, and a land entitlement of one hectare.

In all three states, homestead plots of 500 sq m (5400 sq feet) are being given to all those affected and to their major sons, and to major unmarried daughters as well in Maharashtra.

The NWDT had in addition prescribed a resettlement grant of

Rs. 750 and a grant-in-aid of up to Rs. 500 per family. As against this, Gujarat and Maharashtra are offering a subsistence allowance of Rs. 15 per day for 300 days spread over a year after resettlement (or Rs. 4500 in all). Further, while Maharashtra will give the prescribed Rs. 500 grant-in-aid, Gujarat has said it will give Rs. 750 per family, increased by eight per cent per annum, with 1980 as the base year, plus the prescribed Rs. 500 grant-in-aid plus a sum of Rs. 5,000 for the purchase of productive assets. Additionally, each family is to receive Rs. 10,000 for construction of a pucca plinth for its residential dwelling. M.P., as stated earlier, is offering a rehabilitation grant of Rs. 11,000 to SC, ST, small and marginal farmers, and Rs. 5,500 to other landowners. Landless SC/ST families and agricultural labourers and their major sons are to receive a special economic package of Rs. 40,000 for the purchase of productive assets, the sum being reduced to Rs. 25,000 in the case of traders, artisans and other landless families. Transportation expenses to the new relocation site will be met in all three states and compensation for lands, houses and other assets (trees, wells, etc) in accordance with the LA Act and at replacement value for houses in M.P.

The NWDT laid down a scale of civic amenities to be provided at the relocation sites in relation to the number of settlers. These include a drinking water well and trough and a tree platform for every 50 families; a primary school for every 100 families; and a dispensary, seed store, panchayat ghar, children's park, and place of worship for every 500 families; plus access roads, electric and water supply and sanitary arrangements. All the states have guaranteed this package and have gone beyond it by including items like cattlesheds, threshing floors, cultural centres, cremation/burial grounds, drains, pasture/grazing land, a middle school if one existed at the old habitation, bus stand, and public latrines (Maharashtra). In addition, employment is to be reserved in the project and elsewhere and all on-going welfare schemes are being extended to the relocation sites. (NCA, June 1993)

The question has been asked as to what happens to these provisions in case project affected persons (PAPs) are scattered or fragmented either by necessity or choice. The State governments have responded by stating that essential amenities will be ensured while in other cases account will be taken of the facilities available in the host villages and those amenities expanded or upgraded as necessary.

Gujarat had offered to make the allocated farmlands fit for cultivation or else give PAPs a grant of Rs. 600 to plough or tractorise their fields so that agricultural operations get off to a good start. It plans to ensure that all families benefit from its rural electrification programme and are assisted to electrify 1.5 points in their homes.

Taken together, Gujarat's R & R package at current prices is estimated to cost over Rs. 1.5 lakhs per family. Given an identified 4,500 beneficiaries in 19 submergence villages, the overall R & R cost would work out to around Rs. 675 crores apart from overheads. The figures of losses and gains are even more striking. 981 landholders will be losing 4,700 acres of private land. As against this, 1404 persons including co-parceners and major sons will be getting 7020 acres of land with each landholder entitled to a minimum of five acres (two hectares). As against 1,000 acres of land encroached on by 803 tribals in the 19 submergence villages in Gujarat, 664 encroachers and their major sons had by May 1993 already got 3330 acres of land. Until that same date, 143 landless agricultural families and their major sons had received 715 acres at the new sites. Of the 4,500 identified beneficiaries, 4164 had been given 20,820 ha of land costing Rs. 24 crores. About 2,500 houses will be going under submergence. As against this, 4,500 homestead plots are being allotted free of cost, each of 500 sq.m. Of the 19 affected villages, only one had electricity which lit a few houses. In the new village sites, practically every house already has electricity. These figures do not include the benefits going to PAPs from Maharashtra and M.P. opting to resettle in Gujarat. (Sardar Sarovar Punarvasavat Agency, July 1993)

As far as Gujarat and Maharashtra are concerned the entire submergence area is rural. Not so in M.P. where the Indira Sagar Project will submerge the tehsil town of Harsud in Khandwa district. Since the R & R package for the SSP and Narmada Sagar Complex in M.P. is similar, the State has set out norms for the rehabilitation of urban oustees in new townships in four plot sizes of 95, 140, 280 and 420 sq. m, depending on their income. Urban encroachers on Government land as of April 13, 1987 will be entitled to 50 sq. m housing plots in the new township on payment (unless they are below the poverty line). Compensation for encroached houses or land will be paid at replacement cost but no compensation will be paid for the land. Housing loans will be sought to be provided. Those losing shops and other commercial property will be given plots at the relocation townships or accorded preference in setting up enterprises in proposed new industrial areas. (NVDA, 1992)

R & R in Gujarat

The issues of resettlement and rehabilitation in Gujarat falls into three parts. Displacement first occurred not in relation to the SSP proper but with the commencement of work on the original Navagam dam (in the same general area) in 1961 at six villages around Kevadia and later at five other villages where a series of depressions had to be dyked appurtenant to the main dam to follow. Displacement from these 11 villages was phased in stages, first in 1961 itself, then as a result of preliminary works following the NWDT Award in 1981 and thereafter with the growing tempo of construction on the main dam after 1985-87. Each of these phases was covered by a different set of gradually evolving R & R norms, the earliest displacement clearly belonging to the ancien regime. Successive changes not only bred discontent among the earlier oustees, who felt comparatively disadvantaged, but created another problem. Efforts to meet these grievances half way (by the subsequent inclusion of major sons, landless and so on) resulted in splitting families or shifting already settled oustees to other areas where they could be united. The IRM and other critics have made great play of this historical fallout which should really be seen as a separate issue to be dealt with on its own merits. The overall criterion remains the same in each case, namely, that all oustees be as well or better off at the end of the day. The Gujarat government has bent its efforts in that direction and cannot be said to have remained indifferent to the plight of the Kevadia-dyke villages.

ARCH took up the cause and, with the World Bank's intervention as well, some improvements resulted over and above the job offers on the dam, canal and other works which had already been availed of. The 1961 land compensation rates were said to vary between Rs. 90 and Rs. 250 per acre. By 1981-85 compensation rates appear to have risen to Rs. 7,000 per acre whereas land values in 1992 were around Rs. 15,000–25,000 per acre (IRM). On the other hand, the Sardar Sarovar Nahar Nigam maintained that 8,500 workers were employed on the dam by that time, 2,100 as labourers earning Rs. 36 and Rs. 25 per day as minimum wages for skilled and unskilled work respectively. A further 10,000 workers were engaged on the Main Canal. The land values cited by IRM obviously reflect the capital appreciation brought about by the project. Nevertheless, an enhanced package was finally announced by GOG by a resolution dated August 19, 1992. This provided for additional cash assistance to the extent of Rs. 36,000 for residents of the six Kevadia colony villages, exclusively for the purchase of land or productive assets. Major sons as of January 1, 1987 were also eligible for this benefit. In addition, each khatedar was to be given a homestead plot and free transport to the relocation site. Major sons would have a choice of either retaining the jobs they had been given in the Nigam or accepting this offer. The new package is said to have been well received. Retroactive application of the GOG's regular post-1988 SSP R & R package to Kevadia colony oustees would have established an awkward precedent which the Government quite understandably wished to avoid. The dyke-village oustees displaced soon after the announcement of the revised 1988 R & R policy were, however, granted these benefits though this has meant some dispersal of these families by choice.

All sides are agreed that there was a marked change in the quality and tempo of R & R in Gujarat after the 1988 package was put into effect. Earlier problems are being overcome. Fear of land not being available has been shown to be unfounded. Rather than seek to settle persons on uncultivated government lands which require to be broken in, or leave it to unorganised and simple oustees to negotiate the purchase of compensatory farmlands, GOG decided to set up Land Purchase Committees consisting of representatives of the oustees, the NGOs assisting them, the Nigam, local officials and elected representatives. The LPCs negotiate the purchase of lands at sites selected by the oustees on the basis of personal visits and expression of choice, rejecting sites not to their liking. The fears expressed by the NBA and others that even if land was found, the prices would be ruinous and thwart R & R, has proved unfounded. Rather than a figure of Rs. 50,000 per acre bandied about by critics, the average price was around Rs. 13,000 per acre, going up to Rs. 23,000 per acre under an arbitral award for 4000 acres of prime irrigated land at Dabhoi.

Criticism by the IRM of the scatter effect of resettlement with tribal oustees having to undertake a "long cultural journey" have been disproved. The basic socio-cultural unit in tribal villages is the hamlet based on internal diversities as between Tadvis and Bhilalas and sub-groups within them. Therefore, there was never any expectation that villages would choose to relocate en masse. Even if hamlets or sub-groups have been fragmented in the process of relocation, the fact is that this has been a result of their own volition through the exercise of free choice. Thus, some of the conclusions drawn by an early World Bank consultant, Dr. Thayer Scudder, who is pointedly cited by the IRM, NBA and other activist groups, are mistaken and represent neither the social reality of the old villages or current social dynamics. (ARCH, April 1991)

The IRM is disingenuous to pillory the denial of full land benefits (2 ha to the landless, encroachers, major sons, *et al*) by Maharashtra and M.P. and then criticise the outcome of free choice. The IRM argues that in being asked to choose between one parcel of land and another they are being forced to make an economic or agricultural choice at the cost of social and cultural needs. This is perverse reasoning. The IRM insists on choice but is unhappy with the outcome of free choice. Why should it — or the NBA — choose for the tribals and want to freeze them into a rigid mould as though imprisoned in a time-warp? ARCH gets to the heart of the matter when it says that the tribals see the R & R package and the relocation sites they themselves select as a "development opportunity". They look forward to economic and social gain without loss of cultural identity as they are relocating in groups in fair proximity to one another and that too in a social and cultural environment which is far from being totally alien. The whole approach of the IRM is anthropologically unsound, smacks of isolationist apartheid, and is gratuitously paternalistic.

The IRM team was taken by ARCH and Anand Niketan to

visit successful R & R villages. The IRM acknowledges the visible benefits which "lead to no clear conclusions"! There is criticism of irrigable lands yet to be irrigated, though irrigation will commence after 1995. But where irrigation has been provided the criticism is that this is leading to cultivation of cash crops (and, others would more pointedly say, to a market economy). Irrigation is a necessary though not a sufficient condition for successful rehabilitation, says the IRM. Agreed. And then: "Much of the oustees' optimism arises not from present benefits (why not ?), but from the belief that promises that have been made will be kept." Which promises made and not kept or are unlikely to be kept are typically not mentioned. This is negativism by innuendo. There are problems of grazing, fuelwood and privacy. These too are being tackled and will go. The IRM, like the NBA, makes the error of confusing transient problems with permanent outcomes.

Scudder Report

The IRM and anti-dam activists make much of the reports of Prof. Thayer Scudder who was engaged as a World Bank consultant in 1983 to evaluate the social impacts of the project. He returned again in 1984 and reported on the relevant absence of data on the number and characteristics of those to be resettled and on shortcomings in the R & R package then on offer. It was ARCH which drew his attention to many of these matters. The World Bank adopted his recommendations and incorporated virtually all of them in its own appraisal and evaluation reports and exchanges with the GOI and other project authorities.

These issues were pursued and steadily corrected in the evolving R & R policies enunciated by the three concerned States over the next few years. By 1989 many improvements had been effected when Prof. Scudder returned briefly as a consultant on a three-man Bank review team. He spent under a week in the project area and cleared the mission's draft report before leaving. However, a somewhat fuller document separately prepared by Prof. Scudder in his individual capacity was leaked to the press and came to be widely cited. Nearly all his points had been taken into account in the joint report of the mission to which he was privy and he certainly did not conclude that resettlement was impossible as suggested by Medha Patkar, the NBA leader. Many of his recommendations were in fact acted upon though some were found impracticable. In any event, by the time the IRM arrived in India the Scudder "report" was completely outdated and those who kept citing it may have taken comfort from recording a fragment of history but were essentially flogging a dead horse. A detailed critique of the IRM's selective and misleading use of Scudder's "report" has been made by ARCH (January, 1993)

A parallel set of recommendations by Oxfam, which made certain critical observations on R & R in December 1989, were also noted by the Bank, the concerned governments and the project authorities. Action was taken on many of these recommendations in due course, or clarifications issued. Oxfam saw no reason for further protest thereafter.

R & R in Maharashtra

In Maharashtra as in M.P. and Gujarat, initial problems regarding eligible categories over and above the NWDT norm were resolved as described. Scattered islands in the reservoir were to be acquired and anxiety over the fate of forest village, as opposed to revenue villages, was overcome by declaring that the former too would be covered by R & R. A major problem remained in the prohibition of diversion of forest lands under the Forest Conservation Act despite a commitment by GOI to permit this in the loan agreements entered into with the World Bank. On the contrary, in granting environmental and forest clearance for the SSP in 1987, the Ministry of Environment and Forests (MOEF) repeated the prohibition and refused to relent despite a series of applications by the project authorities and GOM and reminders by the World Bank that this is what had been agreed. That the MOEF should violate a solemn undertaking made by GOI as a contractual obligation is surprising.

However, the NDS and Medha Patkar, the NBA leader, personally kept demanding the release of degraded forest land adjacent to the submergence area in Maharashtra for purposes of R & R. They identified the Taloda forest as a suitable area and specifically sought its release in a Memorandum submitted to GOM in April 1986. The following year Medha Patkar in a seminar in Bombay sought release of "large chunks of denuded forest land" and lamented that the "forest-tribal linkage too is not viewed seriously or not considered in planning alternatives" (ARCH, April 1991). At another seminar in Delhi she stated that denuded forest land to the extent of 18,000 hectares "available and acceptable to the (SSP) oustees . . . is not being strongly demanded by the State authorities, nor is (it being) released by those at the Centre". (Patkar, December 1987) Bruce Rich of the Environmental Defence Fund of the U.S., cited this as a breach of agreement at a U.S. Congressional hearing on the Narmada projects.

Based on the non-availability of non-forest lands in Maharashtra for R & R and MOEF's refusal to release Taloda forest land, anti-dam critics in India and abroad claimed that rehabilitation was impossible. Finally, in response to a World Bank deadline of compliance by the end of June 1990 in accordance with the terms of its 1985 Loan Agreement, MOEF released 2,700 ha of denuded forest land in Taloda. No sooner was this announced than the NDS made a dramatic about-turn. It denounced this act as reprehensible, violative of MOEF's earlier conditional clearance, and "tantamount to selling India's ecological and social future at the behest of a foreign agency" whose officials had been gheraoed in Delhi only the previous month by anti-dam activists for not insisting on denotification of an even larger area of degraded forest land! The 2,700 ha of Taloda forest land was now said to be grossly insufficient, heavily encroached upon and hence unsuitable and in any case degraded to the point of being non-cultivable. (ARCH)

The NDS went further. It submitted a memorandum to GOM dated July 30, 1990 on behalf of the oustees of the 33 submergence villages in Maharashtra, rejecting the Taloda forest land. (ARCH, November 1992) This was investigated by ARCH as a World Bank team had visited the area in November 1990 and found little encroachment or hostility to resettlement in Taloda. Some months later, Tribal PAPs of SSP in Maharashtra belonging to the Narmada Shikshan Samiti and members of ARCH, wrote a letter to the Chief Minister of Maharashtra through the Additional Collector and Rehabilitation Officer, SSP, in Dhule. They said they had from 1986 been demanding release of Taloda forest lands for their R & R. The land had finally been released in June 1990. They went on to say that they had heard only recently that "the organisation with which we had associated ourselves for our struggle, i.e., the Narmada Dharangast Samiti (NDS), had without consulting or even informing us, and with the support of tribals of non-submerging areas, published within as well as outside the country that we do not want this land, which was damaging to our vital interests of survival. We have never said anywhere that we do not want the Taloda forest lands released because that might damage the environment. In fact, we have been badly cheated. We tribals are illiterate, but the organisation which claims to protect our vital interests shouldn't have given us such a lethal blow. We can now see that the NDS itself was not willing to get the Taloda forest land released for us and this is how the DOE (MOEF) could afford to ignore our demand for two and a half years. These tribal PAPs sought release of additional forest lands to meet their R & R requirements. (Narmada Shikshan Samiti, April 29, 1992) This is a strong indictment of the NDS/NBA.

Another line of attack by the NDS/NBA and other groups was the serious loss of forests involved in the Taloda resettlement option. Medha Patkar sought a Dhule court injunction. Over 500,000 trees were going to be lost "with disastrous consequences on the environment" (Lawyers, 1991). The forests of Taloda were described as a rare green patch in this range of hills and the felling of "such a fine stretch of forests" would be an "irreparable loss". (Alvares, 1991) These estimates of the quality of that part of the Taloda forest designated for R & R are exaggerated as until its release the critics were at the forefront in claiming that these were denuded forest lands, which is in fact the case. In any event, with the diversion of these forests, it is incumbent on Maharashtra to undertake compensatory afforestation elsewhere.

The IRM expressed a different concern. It felt that a very large number of encroachers would be excluded by GOM's R & R policy. This anxiety is misplaced as GOM's Resolutions of June 29, 1989 and February 26, 1992 vested encroachers prior to March 1978 (as surveyed in 1985) with 2 ha of land and post-1978 encroachers with 1 ha of land. With the decision to acquire marooned patches or islands in the reservoir and with forest and revenue villages being treated on par, 1047 out of 1485 PAPs or over 70 per cent of the families in the 24 submergence villages of Akrani taluka would be entitled to 2 ha of land. (ARCH, November 1992)

Other objections about the quality of the land and the ability of PAPs to sustain themselves at Taloda were answered by the Tata Institute of Social Sciences (TISS). Its 14th official Monitoring and Evaluation Report found the overall quality of land and potential yields at Taloda to be superior to those prevailing in the Akkalkuwa and Akrani submergence areas. It found the social and cultural traits of the PAPs and host villages compatible, both being Bhils. However, the PAPs probably enjoyed a slightly better nutritional status than the host population in terms of vegetables, forest produce and fish and care would be required to circumvent any ill health effects in consequence. Likewise, adequate grazing must be ensured to both host and PAP villagers. Taloda enjoyed better access to roads, educational, health and credit and marketing facilities than the submergence villages, but water supply could be a problem in summer without tubewells and hand pumps (which are mandatory in PAP settlements). Indebtedness was somewhat higher in Taloda and some resentment noticed as encroachers living in the area were not being given land like the "outsiders". Community organisers should be appointed to smoothen the transition. (TISS, May 1991)

The IRM's assessment of R & R possibilities in Maharashtra exhibits extreme pessimism which is not warranted by its own findings. In the interior Akrani villages its members spoke to many people "who told us that they are now in favour of resettling. They showed us their fields and said the deforestation followed by soil erosion was drastically reducing the productivity of much of their land". Again, "some of those who spoke to us at Bhusha told us that they had been prominent in the campaign against Sardar Sarovar, but had now come to the conclusion that the R & R process might offer them an opportunity to secure a better agricultural base". However, they all wanted Gujarat terms. But over 70 per cent will be getting Gujarat terms under the upgraded GOM policies which the IRM partially mentions in a footnote and then sets aside. Yet again, the IRM observes that Paura (a Bhil sub-tribe) households showed strong Hindu influence which might explain "why some leading families are inclined to look at R & R as a way of building on social and economic changes that are already taking place". Here is a frank admission of the process of sanskritisation which is denied or viewed with suspicion elsewhere in the IRM Report in another of its many contradictions.

The complaint of submergence villagers lacking information and, therefore, being assailed by uncertainty is partly answered by the IRM which notes that the NBA under Medha Patkar's leadership organised strong resistance to the SSP and R & R in Maharashtra first from Dhule (the district headquarters) and later from Manibeli. "Resistance by villagers to surveyors and other Government officials as well as to Tata Institute field workers, has limited what could be achieved". It also notes the NBA's complete volte face over the release of 2,700 ha of degraded Taloda forest land. The inferences that might follow these observations are not drawn.

Much has been made of reports by TISS and others regarding the position of certain Maharashtra PAPs who opted early on to relocate in Gujarat. They were relocated on government land at Parveta. The initial experience of some of these settlers from the Akkalkuwa village of Manibeli (the first of any expected to be submerged) was not entirely satisfactory. LA notices were served on Manibeli residents in 1981 and payment of compensation began in 1984. Temporary movement to Parveta, 35 km away, started a year later. TISS monitoring and evaluation (M & E) studies commenced some time thereafter. Many families that opted for relocation moved into temporary tin sheds costing Rs. 20,000 apiece while continuing to farm their earlier village lands as allowed by the NWDT Award. The initial findings reflected some of the problems encountered, many of them products of midstream changes in policy which was still evolving, the inexperience of a newly-established R & R organisation and the unfortunate choice of previously uncultivated government land which had to be broken in, alongside normal teething troubles. These are not to be denied and, Parveta being among the first R & R settlements and Manibeli the first submergence village, should have been avoided. Nevertheless, to hold what were transient problems to be evidence of ultimate failure and doom, as the critics did was mistaken. Corrective action followed and things have improved as many PAPs have more recently testified.

The TISS M & E Report No. 16 of 1992 found basic services functioning well at Parveta and keenly appreciated, especially the school. The provision of gobar gas plants had eased the firewood problem for several families. Income from wage labour, at the dam site and from sugarcane harvesting, had buttressed the agricultural economy. "Most of the men and women found the productivity potential and actual crop yield of Parveta land as being superior to the land in Manibeli". (IRM, 1992)

The IRM does not set much store by Maharashtra's emphasis on plans for alternative sources of income generation. Nor does it see prospects of R & R in Taloda improving were MOEF to release another 1500 (or more) hectares of degraded forest land for which GOM has applied. It argues that tribal PAPs lack the skills to enter the job market but does not explore vocational training possibilities. On the contrary it believes that "a shift from farmer/gatherer to landless labourer" will be an "economic disaster". This appears to be no more than a plea to maintain the status quo which the IRM itself found many among those living in the submergence villages disfavoured.

A group from TISS has since summarised the findings of six years of M & E work done by it for the submergence villages in Maharashtra. (TISS Group, August 1993) This is a curious document. It denies any attempt to glorify the tribal situation "as life in these areas is both arduous and rigorous. Instead their well-being is seen from the point of view of the tribals themselves". The Group notes the existence of an integrated, self-sufficient and essentially non-monetised economy based on revenue or encroached forest land and supplemented by the forest and river. Even if this was or is a happy existence, is it sustainable? The Group itself observes that "whenever a family expanded, the members took up additional land, cleared and cultivated it". In other words, the forest is being steadily diverted to agriculture with expanding population and needs. Having said this, the Group notes that diversion of land for R & R is not possible since "forest land cannot be a source of agricultural land for the re-settled, given the magnitude of the displacement caused and depleted forest reserves". The fact is that the forests have already been depleted by the tribals and, if continued unchecked, would lead to its complete disappearance before long. Like the IRM, the TISS group appears to be advocating the status quo. It laments that "a project symbolising development (the SSP) has caused the deprivation of a group of people who lived beyond 'development' strategies for centuries. Bringing people into the mainstream should not adversely affect their life situation". The Group's bias towards the NBA is openly confessed in the statement that "the government is attempting to resettle people by marginalising the advocacy groups spearheading the anti-dam movements". The time has come to look for alternatives to large dams. The NBA has "sown the seeds of assertiveness and it has taken firm roots in the minds of the affected people". It is, therefore, incumbent to review not only R & R but "the entire issue of creating development with deprivation". There would appear to be more than some variance between what looks like an informal TISS document and the formal TISS position. If this is so, then the fact that an official consultant or its staff should publicly flaunt a "private" bias is both revealing and disconcerting.

R & R in Madhya Pradesh

The situation in Madhya Pradesh is somewhat different. R & R policy was slow in evolving but by 1992 had been considerably upgraded above the NWDT norms as earlier indicated. Three factors distinguish R & R pertaining to SSP in M.P. from that in Gujarat and Maharashtra. Submergence occurs much later than in the other two states; PAPs in the upper reaches of the reservoir are physically and culturally more distant from Gujarat and may therefore prefer to resettle in M.P. itself; and, finally, although the number of affected villages in M.P. is far larger than in the two other States, the bulk of the villages are only very partially affected with only government lands coming under submergence in some cases.

The IRM, however, came up with its usual criticism: an R & R package that falls short of that in Gujarat, lack of consultation and data; failure to develop an appropriate organisation; and deep pessimism about the scale of effort required and the outcome. It concluded that of the three riparian States, M.P. was likely to suffer the most severe impacts as submergence would displace twice as many families as in the two other States combined. It saw neither policy instruments nor institutions that would ensure a fair deal to the PAPs.

The Narmada Valley Development Authority (NVDA) set up by GOMP for construction of the NSP complex and R & R of SSP oustees in M.P. was guided by GOI norms relating to forest diversion and tribal resettlement and by the State's Rehabilitation Act of 1985, the M.P. Pariyojna Ke Karan Visthapit Vyakti (Punhsthapan) Adhiniyam, which was to some extent modelled on an earlier Maharashtra enactment of 1976. Land for land was to be the basis of R & R with the creation of land pools and resort to drawdown cultivation during the dry months when the SSP and ISP reservoirs would recede. Employment opportunities, including that in fisheries development, were to be emphasised. (Varma, July 1985)

Subsequent criticism focussed around GOMP's failure to match the Gujarat R & R package. More than that, concern was expressed that M.P. had delayed identifying PAPs in the first 17 to 23 SSP submergence villages in its territory who might wish to resettle in Gujarat. These populations would need to be fully surveyed, enabled to select lands in Gujarat and then relocated in accordance with agreed schedules. All this would necessarily take time and organisation. The World Bank's concerns in this regard were set out in a Memorandum addressed to the Chief Minister of M.P. in December 1988. (Paranjpye, June 1990) Earlier, MARG had found M.P. PAPs relatively uninformed and hence anxious and uncertain about their future. M.P. for its part was hampered by the non-cooperative and even violent attitude of some of the submergence villages towards NVDA surveyors, engineers and other officials. Baba Amte's arrival on the scene stiffened opposition. Despite these handicaps, the NVDA completed a 64-point household survey based on discussions with knowledgeable officials and selected community leaders which went into the preparation of M.P.'s 1992 Action Plan for R & R. Though in the circumstances this might not have been accurate to the last detail it provided a broad approximation for planning purposes. The IRM was, however, utterly dismissive of this effort although the major contours of action were reasonably well defined.

The critics and IRM by simple extrapolation assumed that, given the dimensions of R & R in Gujarat and Maharashtra, the problem in M.P. must necessarily be thrice as large in view of the fact that three times as many villages/families were likely to be affected. They were victims of a powerful muddle as they ignored the very different nature of submergence characteristics even though this was pointed out by the NVDA and the M & E agency for M.P., the Hari Singh Gaur University, Sagar. In the earlier stages, even M.P. officials were not sure of the actual extent of farmlands that would be lost and a figure of 24,000 ha was assumed. (Alvares and Billorey, 1988) The facts were soon established and proved otherwise. There was also an exaggerated notion of the number of encroachers and major sons who might be excluded. These fears too were shown to be unfounded on fuller examination and with the evolution of GOMP's R & R package.

As far as major sons are concerned, many if not most are recorded as co-sharers of landed property with the result that they are treated as landholders and entitled to land as compensation. Again, encroachers recorded prior to April 13, 1987 are entitled to compensation for all encroached lands coming under submergence up to a maximum of 2 ha and a minimum of 1 ha. IRM's apprehension that 60 per cent of the tribal oustees in M.P. will not receive land by way of compensation is therefore wild conjecture. The State government has in fact assessed that 76 per cent of the tribal oustees are recorded as cultivators. The IRM's clinching argument that tribal encroachers would not be able to produce proof of prior cultivation in the form of receipts of fines paid to forest guards is invalidated by the fact that even if tribal families were to lose these bits of paper, as well they might, counterfoils are available with the Forest Department.

What are the facts? Of the 193 M.P. villages coming under SSP submergence, only one, Jalsindhi in Alirajpur tehsil, will be affected in June 1994, 16 in 1995, 13 in 1996, and 163 in 1997. Of these last 163 villages only one, Rohana, in Kukshi tehsil of Dhar district, will be completely submerged. Abadi (village sites) will be fully submerged in 39 villages and partially in 116, with less than 20 ha of farmland being submerged in them. Agricultural land will be affected up to 10 per cent in 82 villages, between 11 and 25 per cent in 32 villages and between 26 and 50 per cent in 30 villages. In 21 villages only abadi will be affected and only government land in another 9 villages. So submergence is largely partial. The total land affected will be 20,722 ha, of which no more than 6598 ha is private agricultural land, 2731 is forest land and as much as 11,393 "other government land". (NVDA, March 1993)

The total number of affected families is now estimated at 33,014, as many as half of these being major sons. The landed families number 19,725 and the Scheduled Tribes among the PAPs aggregate 29 per cent. Until March 1993, 8180 families had consented to settle in Gujarat while an almost equivalent number had yet to make up their minds. As many of these families are in areas nearer Gujarat, with which they have socio-cultural affinities, it is assumed by GOMP that a majority might finally opt for Gujarat, bringing the total number of families wanting to relocate in Gujarat to around 13,100. This would be an upper limit. Of these some 1,360 families had already shifted to Gujarat by June 1993.

These figures suggest a larger movement to Gujarat than originally anticipated but physical and cultural proximity and the advantage of the Gujarat R & R package are factors that would weigh in the decision. A case study of Kakrana village in Alirajpur tehsil is illustrative. This village is inhabited by 395 PAPs of which landholders, including encroachers, number only 94. But they have 144 major sons. Thus, against a submergence of 211 ha, the Kakrana PAPs on relocation in Gujarat will receive about 800 ha of irrigable land. (K.Z. Patel, 1992)

The affected 193 M.P. villages farm just under 50,000 ha of land. Of this only around 13.7 per cent will be submerged (including flooding by backwater effect) whereas out of 21,231 houses in the affected villages, as many as 73.3 per cent are affected. (ARCH, April, 1991) This is interesting and indicates that abadi lands are located nearer the waterfront and, especially upstream, yield a pattern of high housing but little farmland loss. This will be so in the fertile Nimad tract where compensation will largely entail relocating the abadi or affected houses on higher ground. The loss of agricultural land will be relatively small and could be largely compensated by permitting drawdown cultivation along the gentler exposed slopes of the reservoir as the water recedes after November. The IRM has made much of the huge loss of rich agricultural lands in Nimad where Kanbi-Patidar migrants from Gujarat settled in the last century. With lift irrigation along a five km strip on either side of the river, they have attained considerable prosperity by cultivating cash crops through modern agriculture. Small tribal farmers and landless labourers constitute the rest of the community and many households depend on fishing. The IRM laments the impending disruption of these social and economic ties. The anxiety is perhaps excessive as those merely re-siting their homes on higher ground will probably not wish to relocate in altogether new areas or in another State.

GOMP, like GOM, is anxious to promote self-employment schemes through provision of training and financial assistance to the landless and artisans.

The NVDA commissioned a socio-economic study of the tribal PAPs in the 17 villages coming under submergence in Alirajpur and Kukshi tehsils. A number of interesting facets of tribal life came to light (Dubey, K.C., 1993) The Bhils of this region share the same cultural milieu as their cousins in Gujarat. An impoverished agriculture yields only limited sustenance, compelling the Bhils to migrate for wage labour to Gujarat for about five months in the year. They are also inveterate tree-cutters and go to the adjacent Maharashtra forests to fell trees illicitly for sale. Dependence on the forests in their own habitat is only nominal as there is no forest left in certain areas. Land records need to be brought up to date to define more accurately the situation regarding encroachers and co-sharers in the land. Some of the SSP-submergence tribals in M.P. who are not listed as Scheduled Tribes in Gujarat, are to be included as Scheduled Tribes in that State. Various rituals must be performed on relocation to ensure compliance with custom, and "informal participation" evoked through consultation with traditional tribal leaders such as village elders, the witch doctor or medicine man, and the village head.

The position in M.P. is now very much better than before. The household survey record has been updated, circumventing anti-dam activists where necessary, and each PAP has been given a "submergence card" giving full data about the house, land and property coming under submergence. After LA notifications are issued, each PAP gets an identity card. PAPs are thereafter given three choices for relocation, consent is taken in writing and the lands selected purchased either through land purchase committees or directly (in which case registration fees are exempted). Compensation rates for land, housing, trees and other assets have been laid down and PAPs will be paid 30 per cent solatium on the sum for involuntary acquisition and 12 per cent interest until payment is actually made. Resettlement sites in M.P. have been identified and are being developed. 1228 ha of land have been located in Jhabua district while a number of families have relocated in Gujarat. About 3000 to 4000 PAP families will have to move to Gujarat each year over the next three years to keep pace with the submergence schedule. The task will not be easy, but neither will it be "impossible" as the IRM and the NBA have forecast.

Anti-Dam Struggle

The movement against the Narmada dams has aroused deep emotions. Until 1987-88 the major thrust was on the impossibility of satisfactory rehabilitation in view of what were seen as shortcomings in the Gujarat R & R package and the ban on diversion of forest land in Maharashtra. The lack of momentum in M.P. was confounded by the resignation of the NVDA Chairman, S.C. Varma on the ground that there appeared to be a lack of political will to fulfil the promises made and implement the R & R programme in earnest.

The announcement of the revised Gujarat R & R package met all the earlier objections and disproved the activists' major premise. The focus thereupon underwent drastic change. Environmental objections were pushed to the fore and an ecological disaster was forecast. Steady improvement in the GOM and GOMP R & R packages followed. The subsequent release of degraded forest land for R & R in Maharashtra came as a shock to the NBA which made a smart somersault denouncing the very policy decision it had vigorously berated the Government for not conceding earlier as the only way to ensure satisfactory rehabilitation. The slow percolation into India of Goldsmith and Hilyard's

"The Social and Environmental Effects of Large Dams" (1984) and the statements and actions of MOEF had encouraged a certain environmental radicalism. "Rehabilitation is impossible" got subsumed in a larger "stop-the-dam" paradigm.

The major Indian Narmada-related NGOs divided. ARCH-Vahini, Anand Niketan and others who had spearheaded the movement for better R & R decided, their battle won, to fall behind the project and help ensure its effective implementation, acting as watchdogs of PAP interests. The NBA, which had by now had come into being, absorbing the NDS and other formations under Medha Patkar, and several other groups decided to go the other way, their focus turning on environmental issues, economic viability, tribal welfare, human rights, and alternatives to SSP. All this now seemed part of a new thrust to chart alternative development strategies generally.

It was at this juncture that the Environment Defence Fund (Lori Udall) of the U.S., OXFAM (John Clark) and Survival International (Marcus Colchester) addressed an urgent letter to Anil Patel of ARCH and Medha Patkar of NDS. They felt the international campaign against SSP was at a "critical stage" regarding R & R as the World Bank too was beginning to feel the heat and planning a major review through the then Vice-President, Mr. Moeen Qureshi. Should the fragile unity of the NGOs on the R & R issue break on this point, all the ground gained could be lost, they pleaded. "NDS may wish to go further to take up broader concerns about the desirability of the whole SSP project but is unlikely to want to lose the bargaining power it has achieved on the R & R issue . . .". They urged a consensus on concerns about environmental and other factors beyond R & R regarding which there were still a number of outstanding issues such as "self-determination" for oustees in their R & R choices, access to denuded forest land, secondary displacement and so on. They praised the two NGOs for shaking the World Bank and closed with the appeal that "we need you to give us a united lead". (October 1988) ARCH declined the honour of having foreign NGOs with their own agenda fire their salvoes resting their guns on its shoulders.

Late in 1988 over 300 citizens, including many well known persons, signed a memorandum urging a thorough reappraisal of the Narmada projects. This was forwarded to the Prime Minister by Baba Amte. In the months that followed, the Official Secrets Act was invoked to pre-empt a large protest demonstration at the SSP dam site. (Amte, July 1989) In September 1989, thousands demonstrated at Harsud (in the ISP submergence area) in a rally against the Narmada projects. In March 1990, there was a *rasta roko* (traffic stoppage) on the Bombay-Delhi highway and a couple of months later a *dharna* (protest sit-in) of oustees and activists outside the Prime Minister's residence in Delhi.

The foiling of a Harsud anniversary rally at Kasrawad in M.P. in September 1990 signalled preparation for a *sangharsh yatra* (a long march) from Badwani near Baba Amte's camp to the SSP dam site in a bid to stop further construction. The march started on December 25, 1990 and activists and oustees, including some tribal PAPs, took an oath by the Narmada not to call off the yatra until their goal was achieved. The march was stopped just short of the Gujarat border as thousands of outraged Gujaratis rallied at Ferkuva in defence of the SSP for which they said they had waited for decades. Section 144 (prohibiting an assembly of more than five persons) was declared in the area to prevent any untoward happening should the two rallies converge. Baba Amte crossed the border but encountered the Gujarati dharna. Medha Patkar and six others resorted to fast. GOG agreed to meet the agitationists, but nothing came of it as the NBA insisted on the stoppage of construction work on the dam as a pre-condition. This was unacceptable. The Gujarat government was unmoved by the fast which was called off after over a fortnight towards the end of January 1991. Baba Amte and the NBA pledged to continue the struggle.

The World Bank President, Babar Conable, had by now decided to appoint an independent mission to review the SSP. This was announced. The IRM started its work that September.

There was announcement of the formation of a Citizen's Committee under Justice Krishna Iyer, T.N. Chaturvedi, former Comptroller and Auditor-General, and Prof. Yash Pal, Chairman of the UGC, to review the issues of R & R and environment. Little more was heard of it.

Do or Die at Manibeli

The agitation continued at various levels. So did construction. R & R, particularly in Gujarat, began to gather momentum and, with the new policies coming into place, in Maharashtra and M.P. as well. The dam was steadily rising and seemed to send out an inexorable message. Although no inundation of any village was expected until 1993, a possible one-in-a-hundred-years flood could have caused a backwater effect sufficient to flood the lower habited slopes of Manibeli and the Shoolpaneshwar Temple in Maharashtra temporarily during the 1992 monsoon. This being the first village to face submergence in the entire Narmada valley, Manibeli acquired a mystic significance. Relocation of this mixed Tadvi-Bhilala village had commenced in 1984-85 and should the water rise everybody might leave. The NBA was determined to make this a test case for, if Manibeli was vacated, all might be lost.

The NBA set up camp at 'Narmadai', a typical Manibeli dwelling beside the Temple, determined to prevent any further movement from the village. The district court at Dhule was petitioned. Project authorities and officials were kept at bay.

Under the NWDT Award and administrative procedures agreed upon all round, no submergence is to be permitted until the affected PAPs have been relocated. Under the Tribunal Award, submergence areas must be evacuated at least six months before inundation. The Maharashtra government accordingly issued notices to residents of Manibeli and eight other villages in Maharashtra towards the end of 1991 for relocation of the remaining families by January 31, 1992. The NBA secured a stay and on its vacation in March, officials moved to effect relocation of those who had given their consent in writing and to enable those who had temporarily moved into tin sheds at Parveta and other relocation sites in Gujarat to return and dismantle their Manibeli homes and transport and erect them in their new places of residence. This was sought to be impeded by anti-dam villagers with NBA support, the village dividing between Tadvis who wished to relocate or had already done so, and Vasavas (Bhilalas) who occupied higher ground. Police moved in to protect those wishing to dismantle their houses and move. The NBA propagandised this as resort to intimidation and repression and some sensational media reports appeared.

The NBA was on weak ground in crying foul. The Dhule court which vacated its stay on March 3, 1992 found that of 267 PAPs which the plaintiffs claimed to represent and on whom final eviction notices had been served, 153 had already accepted lands in resettlement sites in Gujarat and Maharashtra and their properties in Manibeli had long since been acquired. All that remained was for them to dismantle their homes in Manibeli which they had been permitted to keep till final eviction so as to enable them to cultivate lands at both places. Regarding the others, the court found that they had declined to exercise their option for relocation at sites of their choice although the project authorities were prepared and able to offer homestead plots, civic amenities and irrigable lands. The Court further found that the plaintiffs had not sought the permission of these 114 to represent them and were therefore incompetent to do so. The Dhule judge cited an earlier order of the Supreme Court on a petition filed by Dr B.D. Sharma, then Commissioner for SC and ST, precluding any stay on the SSP. The Order in fact stated that work should proceed expeditiously even as R & R was monitored by a high level committee headed by the Secretary, Department of Social Welfare, which was required to submit periodic reports to the Court.

The NBA's suit was dismissed. What stands out is the Court's finding that the organisation had resorted to subterfuge to claim a representational status in this matter which it simply did not possess. (Dhule Court, March 3, 1992)

Unfazed by this order, the NBA determined to continue its struggle. In view of allegations of police repression, the Union Water Resources Minister, V.C. Shukla appointed a fact-finding committee to inquire into events in and around Manibeli between March 26 and April 4, 1992. The team comprised representatives of the three SSP Monitoring and Evaluation agencies, namely, TISS, Bombay, the Centre for Social Studies, Surat, and the HSG University, Sagar. Dar Sarati Acharya of TISS was appointed coordinator. The two other members were representatives of the Ministry of Welfare and the NCA. The committee screened video-tapes, took evidence from officials and residents at Manibeli and PAPs who had shifted to Parveta. It found that genuine PAPs had voluntarily moved. No violence was used nor did the police use force of any kind. No atrocity as alleged was reported, and none unwilling to move had been forced to relocate. In other words, the NBA's allegations were dismissed. (Fact Finding Team Report, 1992)

However, another five-member inquiry committee headed by former Justice H. Suresh of the Bombay High Court, Mrinal Gore and three others was said to have concluded that there was large-scale police repression and atrocities on innocent villagers of Manibeli (*Times of India,* April 1992). ARCH-Vahini, whose volunteers assisted the Manibeli residents to shift, issued a rejoinder stating that the team led by Justice Suresh did not meet it or any other pro-dam NGO or Gujarat PAPs who had initiated the voluntary shifting of Manibeli residents. Nor did the team visit the Manibeli PAPs at Parveta where they had relocated. ARCH accordingly questioned the credibility of the report.

Fortunately, the 1992 floods were not severe and, despite the fact that not all families at risk had moved, nothing untoward happened. By the summer of 1993, however, the dam had risen to a height of 61 m at its lowest point in the middle. Although the construction sluices at 18 m had not been closed as planned, it was expected that parts of Manibeli would be submerged on account of the backwater effect. The NBA had adopted the slogan that the villagers would drown rather than relocate and, as the monsoon approached, volunteers and tribals from neighbouring areas began to converge on Manibeli under the activists' banner.

In March 1993, the Government of India announced that it would stop further drawals on World Bank credit for the SSP but would continue with the project on its own. The NBA countered with a *lok nivada* or people's referendum in the course of which over 22,000 homes in 220 villages in Gujarat, Maharashtra and M.P. are said to have put up metal plaques with the legend "we will drown but not move". Police were deployed to prevent demonstrators marching on Kevadia and Manibeli during May and Sec. 144 was imposed in the affected Maharashtra villages.

Defiance, arrests and dharnas followed and Medha Patkar and some tribals started an indefinite hunger strike in Bombay. The Gujarat High Court granted a stay against the forcible eviction of affected persons from any of the 13 Akkalkuwa taluka villages in Maharashtra, including Manibeli, until they could be fully resettled in accordance with the established norms. The Shoolpaneshwar deity was ceremonially shifted to a new temple below the dam site. The Bombay High Court upheld the Dhule SP's order prohibiting persons from the 36 Maharashtra villages from taking part in any *jal samadhi* (drowning) *andolan* at Manibeli or elsewhere. The Supreme Court declined to intervene. Baba Amte staged a fast in Delhi and Medha Patkar broke hers following GOI's offer to enter into a dialogue on June 18. However, the terms of reference continued to be disputed and it was not until August 5 that the NBA, after much drama, mediatory efforts and other interventions announced withdrawal of its *jal samarpan* or mass drowning campaign.

Baba Amte was critical of Medha Patkar's fast unto death which he said was an "obsolete concept" but he favoured resumption of the *jal samarpan* programme if a review of the SSP was not remitted to a core group as decided at the dialogue the Minister of Water Resources conducted with the NBA and other observers at the end of July, 1993. Many were critical of the fast-unto-death and *jal samarpan*. The person whom Medha Patkar said had brought her into the Narmada movement said "leaders have no moral right to lead their followers to destruction . . . Both from the point of ethical and political criteria, the strategy of the Samarpan Dal does not pass muster. . . Collective samparan is collective suicide". (Dhagamvar Vasudha, September 1993)

The Government finally announced the formation of an Eminent Persons Group in August to hear the NBA's views and contentions concerning all aspects of the SSP. The group consisted of Dr Jayant Patil, Planning Commission member, and four others unconnected with the government or the project. There appeared to be considerable confusion with regard to its precise role. The NBA desired and understood the purpose of the Group to be to "review" the project in its entirety. The Minister for Water Resources, however, told Parliament that while matters pertaining to resettlement and environmental impacts could be discussed, the basic parameters of the SSP project as set out in the NWDT Award were legally binding and could only be changed with the consent of all the concerned parties. The Group was expected to complete its work within three months and its report was to be published within a month thereafter. Work on the dam and canals meanwhile continued. The Government said it would consider the views of the Group. The NBA reserved its position, stating that it would determine its future course of action after seeing the Report and, presumably, GOI's response to it. Gujarat was strongly opposed to the setting up of the Group and boycotted the proceedings. The matter went to the Gujarat High Court. The Committee soldiered on.

In the Name of the Tribals

Resettlement and environmental issues are at the fore of the Narmada controversy. But at the core of the R & R issue is the fate of the tribals. The IRM espoused their cause as much as the anti-dam activists, none, however, more ardently than Dr. B.D. Sharma, former Commissioner for Scheduled Castes and Scheduled Tribes, a constitutional office. Dr. Sharma was moved to write about the plight of tribals generally and the fate of those living in the submergence villages of SSP in the 29th Report of the Commissioner for SC & ST which was presented to the President on May 29, 1990. He saw the problem as the tribals' "right to life with dignity", "their access and command over natural resources in their habitat and the serious dissonance between the law and the social customs of these people and their economic systems which has resulted in 'criminalisation' of whole communities as perceived from the high pedestals of the modern state". With large development projects "the tribal people face destruction as individuals and the spectre of disorganisation as communities" This was the principal plea in his subsequent writ petition before the Supreme Court against the Sardar Sarovar Nahar Nigam Ltd. of Gujarat. (Supreme Court, 1990) His prayer to the Court was that it direct that work on the SSP be stayed until the basic issues raised by him were settled "and the concurrence of the tribal people is obtained to the acqusition of land and to the plan for their alternative settlement, if any, based on full facts, clear understanding and free will of the concerned people".

Dr. Sharma raised a human rights issue as "rehabilitation was not possible". Could people be evicted from their homes without their consent? Was the social identity of a person negotiable? Could justice be determined on the basis of the arithmetic of numbers? Could there by a barter deal between the survival needs of some and more facilities and greater comforts for others?

The GOI's counter-affidavit said development was intended to benefit the people and especially the tribals and other weaker sections; a liberal R & R package had been evolved; the basic approach was "to convert the dislocation into a potentially beneficial opportunity for economic betterment"; all development entails some initial disturbance and if this is not allowed then all progress must come to a standstill; the right to life must include the right to development, the right to prosperity and the right to a better standard of life.

The Supreme Court delivered its judgement in August 1991. It found that R & R was making progress and ordered that rehabilitation should be complete at least six months before any area was submerged. With regard to the first phase of submergence, it required the NCA's R & R Committee headed by the Secretary, Ministry of Welfare, Government of India, to supervise progress and submit monthly reports to the Court. In making this order the Court observed that "it is not our intention to hold up the progress of the work. On the other hand we would like it to be completed expeditiously". (Supreme Court, 1991)

The IRM devotes a whole chapter to "Tribal people and the Valley" and attempts a definition of who is a tribal. This is a laboured exercise as there is no ambiguity about who is a tribal and entitled to tribal status in India, this being defined by formal state-wise notifications in accordance with constitutional provisions. The distinction between forest-dwellers and "forest villages" and tribals is also not material to defining tribes and nothing turns on it. Virtually all the tribes in the lower Narmada valley affected by SSP are part of the Bhil family which numbers several million and is spread over M.P., Maharashtra, Gujarat and Rajasthan. In any event, the IRM had no mandate whatever to foist its own definitions and interpretations on a country where these issues have been constitutionally governed without the slightest controversy or misunderstanding since 1947 and even prior to Independence. Other weaker sections, though not tribal, also receive constitutional protection. The IRM itself notes that the World Bank's loan agreements on SSP make no reference to tribals. From this it deduces that the Bank did not know of or ignored the presence of tribals in the SSP project area which it, IRM, had dramatically uncovered. Hence its observation: "... lack of basic information about the peoples of the submergence area has contributed to uncertainty within the Bank about the tribal issue. And it has allowed those who have wished to minimise the issue to assert, on the narrow basis of the findings of the 1987 (Bank) mission, that the tribal population of the submergence area is limited to a very small number of extremely isolated villages".

The very idea that the Gujarat government or the Indian authorities have tried to reduce the number of tribal PAPs by failure adequately to define who are tribals, is absurd and betrays a total lack of understanding of the Indian constitutional and political system. Hence also the IRM's muddled references to tribals being passed off as "backward Hindus" and to the process of sanskritisation. All this might sound profound to innocent audiences abroad but will not be taken seriously by Indians. The IRM has tried to make a mystery out of nothing to argue a case which it is unable to bring to any kind of conclusion since there is none to be reached. The only logic that emerges is an effort to argue a distinction between "tribals" and "Indians". It is in this vein that it refers to "the distinctive cultural heritage of these people, the isolation in which they were preserved, and their long history of resistance to external administration". It recalls the World Bank's use of the phrase "semi-tribal" in some of its general operational guidelines and states that "concern for such groups is an aspect of the world's increased awareness of how isolated cultures have all too often paid an appalling price for development. The mechanisms by which they have been separated from their lands and stripped of their own cultural integrity are all too well known". It is not clear what this is meant to suggest. But it seems suspiciously like an effort to wash away latter-day post-Columbian guilt in the waters of the Narmada. If so, it will not wash.

The IRM, like many Indian activist groups, also equates Narmada Valley R & R with human rights. ILO Convention 107, adopted in 1957 and ratified by India is cited. This relates to the Protection and Integration of Indigenous and other Tribal and

Semi-Tribal Populations in Independent Countries. The Convention states that the right of ownership, collective or individual, over the lands which these (indigenous, tribal or semi-tribal) populations traditionally occupy shall be recognised. Article 12 provides that "these populations shall not be removed without their free consent from their habitual territories except in accordance with national laws and regulations for reasons relating to national security, or in the interests of national economic development or the health of the said populations". Removal, where necessary, must be compensated by provision of lands of quality at least equal to that of the lands previously occupied by them, "suitable to provide for their present needs and future development".

The IRM adopts the view held by Indian activist groups that tribals are owners of the forests and have been wrongly labelled encroachers following declaration of their lands as reserved and protected forests by the Government in the recent past. Encroachers have been periodically recognised and regularised and have not been left without benefit or compensation in any of the several SSP R & R packages, with alternative employment and monetary compensation as a supplement or in addition to land in some cases. There has been no breach of ILO Convention 107 in the Narmada Valley.

The U.N. Commission on Human Rights early in 1993 adopted a resolution on forced evictions affirming the right of every individual to a secure place to live in peace and dignity. While its emphasis was principally on housing it does have a wider application. It recommended immediate restitution, compensation and/or appropriate and sufficient alternative accommodation or land, consistent with their wishes and needs, to persons and communities which have been forcibly evicted, following mutually satisfactory negotiations with the affected persons and groups. The displacement in relation to the Narmada Valley projects is not forced eviction in the sense of arbitrary ouster, but is in consonance with law, planned development, and assisted relocation on generous terms, judicially determined, closely supervised and based on choice. Land acquisition for a public purpose is well established and is done in accordance with prescribed procedures which are justiciable. It would be unreasonable to suggest, as some have done, that construction of SSP is based on forced eviction. It is true that some tribal PAPs in Maharashtra and M.P. have thus far been unwilling to relocate and a few threatened *jal samarpan*. It has been alleged that this resistance is the result of instigation by anti-dam activists. However, this is not a material issue even if true.

In any democratic society, governed by the rule of law, individual interests, while precious and subject to strong safeguards, cannot override the larger community or national interest defined through democratic processes. In the case of the SSP, the courts have upheld the continuance of construction and consequent relocation to the extent necessary as lawful and legitimate. In recognition of the social and spiritual trauma and hardship caused by involuntary displacement, the NWDT Award and subsequent government resolutions and orders have enjoined a caring and generous approach on the part of the State towards those affected. They must all be as well if not better-off than before at the end of the process. Governance in any democratic society is based on parliamentary consent and due process within a constitutional frame. Due process here has allowed PAPs to choose their place of relocation. Refusal to exercise their lawful right of option by some cannot be said to amount to "forced eviction" or a human rights violation. Non-construction of the project equally affects the rights and well being of many millions of others. A balance has to be struck and the courts are there to ensure that justice is done.

The special pleadings of Asia Watch, Greenpeace, Friends of the Earth, the Narmada International Human Rights Panel, and other international bodies are possibly well-intentioned but misinformed about the facts and the ground realities. The Human Rights Campaign on Narmada, a consortium of Indian human rights groups, also appears to be labouring under a number of misperceptions. (Asia Watch, June 1992; Mohanty, May 1993) Many charges have been made from time to time of PAPs, including tribals, and anti-dam activists being subjected to various forms of indignities and police repression. While cases of abuse of authority are to be deplored in any circumstances, many allegations such as those surrounding the Manibeli agitation and the *jal samarpan* movement stem from activist-supported obstruction. There has been intimidation of those willing to move or those who have moved, as well as of project surveyors, engineers and other administrative staff, contractors' personnel and government officials. There has been both passive and active violence and the IRM and others have recorded this. There have been cases of assault. Manibeli PAPs, who temporarily relocated at Parveta, complained of social boycott and denial of their customary rights in the village whenever they returned before finally shifting for good. Having prevented surveys and service of notices, the NBA and other activist groups have subsequently made a grievance of inadequate surveys, studies, consultations and so on. Those offering satyagraha and going on indefinite fasts in the name of Gandhian struggle have shown little regard for Gandhian values in not accepting the consequences of their action.

Many of these attitudes and misperceptions stem from the romanticisation and mythification of tribal life in the minds of "outsiders", Indian and foreign, who are disenchanted with the present and are in quest of the promised land which they think they glimpse in the frozen past of present-day tribal life. The IRM and several activist groups recall a sylvan forest setting, the hills and river, and picture a happy symbiotic relationship between the Narmada Valley tribes and nature. They evoke the image of almost enviable hunter-gatherers whose wants, unsustained by an unviable agriculture, are met by foraging for roots and tubers in the forest and fish in the river. The reality is pathetically different, harsh and often grim.

The World Bank tumbled on this truth early. A hand-out (April 1990) noted the widespread myth that the tribal peoples to be resettled from the submergence area of SSP "are living in pristine forests in a traditional manner as hunter-gatherers in harmony with the environment. This is not correct. As satellite imagery shows and field observation confirms, the tribals mostly farm barren, stony, steep and increasingly degraded hillsides with a small and decreasing part of their income derived from the forest land. Many of them welcome the opportunity to improve their lot in the more fertile and more sustainable command area and, while any move has considerable social costs for a family, many have quite widespread tribal connections in the resettlement areas in Gujarat which is expected to help their adjustment. A substantial portion of tribals appears to see little future in the increasingly degrading submergence area. Indeed, it is within this tribal group that there appears to be the least opposition to the dam. The greatest opposition is from the more wealthy tribal farmers"

Another sociologist, whom the IRM also quotes, is equally emphatic in stating that "the romantic image of the 'tribal' does not match the reality of adivasi existence as variously experienced in the Narmada Valley. (Baviskar, Amita, 1991) Nor is the tribal economy sustainable. During her stay in the Valley, Baviskar was "quite unsuccessful in finding an indigenous notion of sustainable development . . . Adivasis seem to believe that the forest is inexhaustible and this faith in the regenerative powers of nature paradoxically facilitate its destruction"

More detailed evidence of such myth-making is to be found in a study of the first 17 SSP submergence villages in M.P. in the districts of Jhabua, Khargone and Dhar by an NVDA team of researchers. (Dubey, March 1993). In a compendious volume entitled *The Bhil Track*, the team says there is no forest worth the name in these villages and in the area in general "and the dependence of the people on the forest is as mythical as the forest itself". The 17 villages have a land area of 19,240 ha of which only one single hectare is under forest! The "forest lands" have been heavily encroached and there are several co-sharers in each parcel. Agriculture, not the "forests" nor fish, is the mainstay of what is a subsistence economy, and the governing factor here is the severe shortage of cultivated (even culturable) land, all of which has very poor soil. In fact, "the fields are nothing but rubble", many on hill slopes, and very little is irrigated. The average size of holding in the possession of a household ranges from 0.88 to 1.70 ha, with two to eight co-sharers in many cases. Some have shifted to Gujarat, and others wish to go. At Kakrana they said, "making our sweat and blood as one, we do not get enough to eat for the whole year. The land there (in Gujarat) is good and we will go whatever some people may say". There is some opposition to movement too elsewhere. Though deeply attached to their land, these Bhils cannot sustain themselves through agriculture or forest produce which is seasonally limited. Many migrate for five to seven months (sometimes with their wives) to Gujarat where they join work gangs or pick cotton. Others (mostly Naiks, a sub-tribe) raid the forests in adjacent Maharashtra to fell timber or fuelwood which they sell. Such timber-theft sometimes entails seven or eight days return march.

The NVDA research team visited some of the M.P. resettlement sites in Gujarat. They found the PAPs socialised and socialising — a new trend — and enjoying the benefits of better land, more produce, higher incomes, access to medicare and education. Children were clothed and women had purchased ornaments. Wrist watches and transistor radios were to be seen. Higher value crops were being cultivated. Addiction to liquor had waned. But there was a darker side too. For lack of fish and meat (as they did not bring their goats and poultry birds with them) there was some evidence of malnutrition among children and scabies from wearing synthetic garments that are not regularly washed and from their inability to take a bath under hand-pumps which have replaced the flowing river. There are some fuelwood and grazing problems too. But these are early days and with social education, gradual resort to stall-feeding and supplementary diets conditions should improve. However, administrative vigilance is necessary to ensure that these problems are indeed overcome.

The Team has some comments on the IRM report. It has pointed out certain errors. Annual values of seasonal forest produce have been sought to be derived by a simple multiplication of a week or a month's income, by 52 or 12, ignoring the fact that some forest harvests are seasonal and of very limited duration! They scoff at IRM's reference to "heavy dependence" on the forest and the river. It is nominal. Attachment to the land, to the forest and to the Narmada are seen by the NVDA team as generally more Indian than specifically tribal (Bhil) traits. The process of cultural inter-penetration is noted. Isolationism will no longer work. Change is bound to come but this should happen in a manner that does not sweep weaker communities off their feet.

Nevertheless tribals and all peasant people are strongly rooted in the soil and their natural environment everywhere. Yet objective realities cannot be ignored. With the growth of their population and livestock holdings, the traditional tribal economy in the lower Narmada valley, as in other parts of India, cannot sustain the people for the better part of the year in many cases. Distress migration to more prosperous agricultural areas or to construction sites or urban slums in search of wage labour is growing and takes place in the most unsatisfactory conditions. The movement is from the land and farming to non-land, non-farming occupations. This distress migration of what might be termed Malthusian refugees is far more worrying than the displacement of PAPs to planned relocation sites with homesteads, civic amenities, land and employment which offers them a real opportunity in life.

There is a certain callousness and arrogance in attitudes that demand that tribal people be denied a choice and be compelled to live in romanticised misery as social relics. Grubbing for roots in the vanishing forest to keep oneself alive cannot be the concept of a happy life with dignity. Nor to be denied access to primary education and health. Certainly the State must provide and cannot neglect any category of citizens. But the State must provide for all. And if there is objection when it makes a start, where does one begin?

Much has been made of the alleged failure to consult the people about whether they wanted the SSP or relocation. Consultation in the prevailing social setting of hitherto sequestered and illiterate communities is not necessarily to be accomplished by literally consulting everybody individually. This has to be done through wider group or community consultation, sometimes indirectly through traditional or elected representatives, and through established administrative and legal processes. This may not be ideal, but there is no better way. The Narmada projects have been debated for two generations. Elected representatives of the submergence areas sit in panchayati bodies, State legislatures, and in Parliament. The Chief Minister of Gujarat during a critical period of decision-making in relation to the SSP was a tribal person. Newspapers, the electronic media, public meetings and administrative interaction have reinforced this process.

The record of consultation of the NBA, the leading critic, is itself not beyond dispute. Baba Amte told *Sunday* (July 10, 1993): "I am seldom if at all consulted on the major decisions. I am merely informed afterwards". Amita Bavaskar wrote in *Lokayan* (1991): The Andolan "objectifies adivasis as symbols meant to touch a responsive chord in liberal hearts, not as makers of their own history. Adivasis are limited to being bow-and-arrow-bearing bodies, indicators of the Andolan's strength, who are not involved in the decision-making process". Further, "the Andolan takes recourse to the manipulation of adivasi symbols. The *shapath* - pledge not to move from the Valley - has been repeated several times in Andolan public meetings - in the name of Narmada and *juvar mata*. Activists argue that these *shapaths* are indigenous and that we are making creative use of adivasi culture. But authentic *shapaths* or mantras occur in a ritual context. . . In the Andolan programmes, we have instant *shapaths* recited over microphones, shorn of their religious meaning. . .".

Listen to Vasudha Dhagamwar (*Indian Express*, 1993) who heads MARG, which had done the survey "Sardar Sarovar Oustees — What Do They Know?": "We (MARG) had only stopped going (on further survey missions) when the NBA virtually directed the villagers not to let us in. In other words, the NBA was in no way behind the Government in hiding information". The Dhule Court, cited earlier, found the MBA petitioning it on behalf of those who had no grievance and others whom it had no title to represent. (Dhule District Court, March 1992) And Akkalkuwa villagers in Maharashtra had to petition the Chief Minister complaining that the fact of the official release of Taloda forest land for R & R had been suppressed by the organisation (NBA) representing them and that it had been falsely reported to the Government that they did not want this land. (Narmada Shikshan Samiti, April 1992) Or again, as reported in a contemporary journal. "Privately (Nalini) Singh narrated an incident. . . She said she was interviewing (Medha) Patkar in Baroda on the same morning that *The Times of India* was screaming Patkar had been arrested. And there was the activist, a free woman, giving an interview to a government medium (Doordarshan). What does this say about the activists' publicity machine"? (Ninan, Sevanti, April 1993)

Canal Oustees

The NWDT Award makes no reference to those losing their land in the SSP command from land acquisition for the canal network. Nor did the Nigam initially provide for anything more than routine compensation under the Land Acquisition Act. The practice has hitherto been to differentiate between the submergence or affected area and the command or benefited area and not to go beyond normal compensation for lands acquired for canals as the requirement is generally modest and those losing land gain greatly from irrigation. However, certain NGOs such as ARCH-Vahini raised this question, following which GOG contracted with the Indian Institute of Management, Ahmedabad to study the matter of compensation to canal affected persons (CAPs).

The IIM report, based on a sample survey, estimated that the construction of 75,050 km of Main, branch, distributary and minor/sub-minor canals would require acquisition of 85,123 ha affecting up to 2,22,786 khatedars (landowners). It felt that the existing method of assessing cash compensation was unsatisfactory and liable to undervaluation. Courts had often enhanced the compensation amount. Moreover, the current compensation formula made no distinction between those losing all their land, those who may be left with under 2 ha and those with more than this amount. (IIM, October 1991)

The Nigam had several objections to the IIM study and informed the IRM that its estimates entailed an "immense overstatement" of the problem owing to the faulty methodology employed. The IRM's assessment led it to conclude that 140,000 khatedars would be affected. Many farmers would be marginalised as a result of land acquisition, approximately 1100 becoming landless and around 7500 being left with less than 2 ha. Though a considerable chunk of the acquired lands would be government-owned, loss of this would affect grazing, fodder and fuelwood to neighbouring villagers. The IRM was of the view that all those rendered marginal or landless farmers should be treated as PAPs and given a minimum of 2 ha each.

The Nigam gathered more data on the canal oustees and, analysing this, arrived at a vastly different conclusion which it presented to the IRM. (SSNNL, 1992) It found the IIM's extrapolation for the entire command, from a sample of 30 out of 165 villages along the Main Canal for which it was provided data, to be very misleading. There are 3393 villages in the command and the quantum of land lost along the branches, distributaries and minors is progressively less. It found that those who lose land in the canal system sometimes hold land at other places in the same or adjoining villages. Allowing for this, only 342 khatedars would be losing all and 936 more than 75 per cent of their land. As it was not the Government's policy to acquire lands for sub-minors, the size of the problem would be reduced to that extent. Only 10.2 per cent of the 152,820 ha owned by land-losers in the command would be acquired. Nearly 146,000 landholders would lose less than 25 per cent of their land and under 16,000 less than half their holding.

The Nigam takes the view that CAPs are not to be compared with PAPs as they do not have to relocate and their socio-economic environment remains unchanged. Further, against the sacrifice of some land through acquisition, farm incomes could triple on their remaining lands with irrigation. In fact, landholders have been very cooperative and anxious to expedite irrigation which will also greatly enhance agricultural employment which is of considerable interest to marginal farmers and landless labourers. Further, where farmers willingly part with their land, obviating acquisition proceedings, GOG is offering 35 per cent over and above the 30 per cent solatium given for involuntary transfer. This amount should enable most farmers to buy land in the neighbourhood. Taking all these factors into account, the Nigam believes that, apart from hard cases for which a special package is being devised, none will be really disadvantaged.

A GOG Resolution of September 15, 1992 set out a package for farmers losing all their lands under canal acquisition or being marginalised. Landholders with 1 ha of land within the command or 2 ha outside it prior to acquisition, and whose entire holding is acquired, will be provided 0.5 ha of land within or 1 ha of land outside the command. Likewise, landholders holding less than 0.5 ha of land within the command or 1 ha outside of it prior to acquisition and from whom part of this holding is acquired, will be provided sufficient land to make up these amounts. The land may be allotted from out of available government or privately purchased land within 8 km of the village where the land is acquired, in accordance with the farmer's preference. This will be at government cost, less the amount of compensation paid for the lands originally acquired.

Overview of R & R

The original, root meaning of the word rehabilitation is "to invest with dignity". That is a truly noble concept and this is what R & R should endeavour to achieve. It is indeed the avowed policy of all the governments involved in the Narmada Valley projects that PAPs should be at least as well if not better off than than they were previously. R & R, however, involves three stages — relocation, which entails physical movement; resettlement, or a process of settling down in terms of housing and livelihood; and, finally, rehabilitation, which is not just a material but an emotional and psychological process of being socially and culturally integrated into the new environment. Rehabilitation is therefore not an instant event but a process that may be spread over some years. The IRM recognises this at one point where it stated that "resettlement and rehabilitation is protracted process. . . The success (of rehabilitation) cannot be judged for many years, possibly not until the next generation". It admits that Gujarat, at least, has demonstrated "considerable success" in implementing the first steps for a large proportion of its oustees. "But rehabilitation is for the most part a set of promises about the future." Why? And from that premise for the IRM to leap to the conclusion that rehabilitation in SSP as a whole is "impossible" strains credulity.

The tendency to view the tribals as isolated and pristine communities is to promote separation and primitivism. The IRM found the remarks of Dr. Vidyut Joshi, formerly at the Centre for Social Studies, Surat, "anything but sympathetic to the protection of tribal culture". Dr. Joshi, discussing tribal culture in the context of larger socio-cultural change, had noted that "a culture based on a lower level of technology and quality of life is bound to give way to a culture with a superior technology and higher quality of life. . . What has happed to us (non-tribal cultures) is bound to happen to them (tribal society) because we both are parts of the same society". Dr. Joshi speaks from experience, having travelled and studied Gujarat's tribal society for 20 years. Hence, "I have formed an opinion that tribals want to change. It seems it is we who have a vested interest in conserving tribal culture for our research, museum and tourism interest". (Joshi, June 1991)

None wants to destroy the finer qualities of tribal life and culture. But to preserve its less happy aspects is another matter when there is opportunity for betterment while preserving tribal identity. This question was widely and long debated in India through the 1950s, especially in connection with the opening up of what was then known as the North East Frontier Agency, NEFA, (now Arunachal), and other parts of the Northeast. Verrier Elwin wrote "A Philosophy for NEFA." Jawaharlal Nehru advocated steering between extremes of isolationism and homogenisation and that is the path India has tried to follow. The process of cultural inter-penetration over centuries is also indubitable (High Level Group, GOG, July 1992) The R & R packages of Gujarat, Maharashtra and M.P. have steadily improved and implementation is well under way with sufficient time before submergence to ensure that there is no "forced eviction". The courts are there to ensure that submergence will not be permitted if R & R has not progressed. NGOs are keeping vigilance. The fears of the NBA have been steadily falsified on the ground and if it were even now to cooperate progress could be faster and smoother.

The critics as a whole have either not understood or have somewhere along the line lost the meaning of rehabilitation. Large generalisations have been made on the basis of limited experience and teething troubles have been assumed to be conclusive and final.

Environmental Concerns

Environmental economics is a relatively new discipline from which has emerged the concept of sustainable development. Environmental costs are often long-delayed, consequences following the triggering action after years or decades to the cost of future generations. These consequences can be remote in time and space, as damming a river can have marked effects both upstream and downstream, along the entire length of the river or over large parts of the basin up to its outfall in the sea. Likewise, many effects can be indirect. Costs may be borne by one group of persons or a region or a generation while the benefits go to another, raising complex issues of equity. All these are valid concerns and there is little dispute about their relevance and importance though there might be debate about their measurement. Individual costs and benefits would need to be viewed against social costs and benefits so that powerful decision-makers are not allowed to pass on costs to weaker subjects.

A fuller understanding of environmental impacts is emerging. Environmental assessments are becoming mandatory and compensatory or ameliorative actions are being internalised in project appraisals and costing. The first World Environment Conference at Stockholm in 1972 was a landmark. India's National Committee on Environmental Planning and Coordination was established that year but the Department of Environment only came into being in 1980. This became the Ministry of Environment and Forests (MOEF) in 1985. Indian and World Bank environmental guidelines evolved through the 1980s and that process continues. What is significant is that the planning, design, sanctioning and implementation of the SSP and NSP Complex followed a trajectory that did not parallel the evolution and refinement of these environmental standards. Engineers and architects always design new machines and structures conservatively, or over-design them as added safety insurance. Redundancy is a well understood concept. Similarly it should be no surprise that in the first shock of environmental consciousness and having to negotiate administratively uncharted waters, governments, bankers and the public might "over-design" environmental norms. This is apparent in some of the environmental controversies surrounding the Narmada Valley projects and the alarms sounded, first by MOEF and the World Bank and then by activist groups and the Independent Review Mission. The IRM, however, was clearly bored by its terms of reference and decided to occupy itself by straying far afield, laying a number of cunning traps along the way and then falling into them. This is not to deny slippages, oversights and delay on the part of one or other of the concerned project authorities or to derogate from the importance of sound and timely environmental appraisals and responsive actions. But it is necessary to place the Narmada environmental debate and the critique of environmental activists in perspective.

Water Availability

Having closed its analysis of the prospects for R & R on a note of extreme pessimism, the IRM introduces its environmental section with the thought that "the Sardar Sarovar Projects (*sic*) will not work as intended" because there is not enough water in the river to deliver the designed results.

The NWDT Award of 1979 determined the heights of the Sardar Sarovar and Narmada Sagar dams, the allocation of waters to M.P., Maharashtra, Gujarat and Rajasthan, and specified the offtake level of the SSP Main Canal on the basis of a balance of water availability and competing claims. At the time the Khosla Committee was engaged in preparing a Narmada Master Plan, discharge observations were only available for the period 1948 to 1962. This being inadequate for any dependable forecast, the Committee hindcast runoff figures from rainfall data recorded since 1891. Accepted rainfall-runoff coefficients were employed and the mean yield at 75 per cent dependability (an Indian norm specifying the quantum guaranteed in 75 out of 100 years) was estimated at 28.92 MAF at Garudeshwar, just below the Sardar Sarovar dam. However, since the rainfall data after 1915 was available from a larger number of stations, it calculated the 1915-62 yields at 75 per cent dependability at 35.94 MAF. This, however, ignored certain preceding drought years.

The matter was argued before the Tribunal whose technical assessors in consultation with the concerned States after careful study agreed to base discharge calculations over a 79 year cycle, 1891 to 1970. In 1974 the four Chief Ministers decided to adopt the figure of 28 MAF thus derived which the Tribunal accepted. This figure was arrived at by taking 27 MAF as the 75 per cent dependable yield, less evaporation losses of 4 MAF, plus regeneration from return flows of 2 MAF and a carryover storage of 3 MAF. Forecasting on the basis of actual data from 1948 to 1970 and later up to 1979, and coupling with this hindcast estimates going back to 1915 or 1891 by various State governments and authorities yielded varying figures. The 1968-70 series gave a meagre flow of 22.50 MAF. The World Bank's 1948-79 estimate placed the 75 per cent dependable flow at 28.57 MAF, that is in excess of the NWDT figure and far above the 25.74 MAF worked out by IIM, Bangalore in its System Studies for Planning of the Narmada Sagar Complex. Yet another expert committee worked out the 75 per cent dependable flow at 30.23 MAF for the period 1891-1949 and at 23.24 for the period 1948-88. (NVDA, September 1991) The variations reflect reliance on different methods of analysis and rainfall-runoff coefficients and different time cycles.

The NVDA committee like the NWDT technical assessors before them, sought to overcome these discrepancies by undertaking a study to ascertain the natural hydrological cycle of the Narmada, taking good and bad years as they come. It took the hindcast figure from 1891 to 1948 and the actually observed flows from 1949 to 1989 and established the hydrological cycle of the river to be 79 years. For any period of 79 years taken between 1891 and 1989, the 75 per cent dependable yield fell between the range of 27.64 MAF and 28.89 MAF, fully confirming the 28 MAF adopted by the NWDT. It was on the basis of this clearer understanding of inflows that the NVDA Committee did not find it necessary to review the Madhya Pradesh Master Plan for the Narmada Valley on hydrological considerations.

The IRM took into account a World Bank analysis indicating that "a 28 per cent error in runoff would reduce the economic return to zero". It then rightly stated that "it is not the responsibility of our review to evaluate the engineering hydrology of the Sardar Sarovar Projects nor is proper for us to review the NWDT's extensive treatment of the issue". But having come across discrepancies in data and analyses, which would affect environmental impacts, "we engaged an independent hydrologist and asked him to test the key matter . . .". Further, "our work on hydrology was designed to better understand impacts. It was not our purpose, nor was it within our Terms of Reference to do a complete systems analysis. But in the course of this work we have found that there is compelling evidence that the Sardar Sarovar Projects will not operate as planned. It is not that the results from the work done in the 1970s and 1980s are wrong. We only know that with more data that is now available, it is possible to reach conclusions that will lead to quite different impacts than have been assumed to date".

The name, terms of reference, methodology, data examined by and findings of the IRM's "independent hydrologist" are not mentioned — and were never subsequently disclosed despite requests. Nor were the NWDT, CWC and World Bank said to be wrong. Yet, the IRM sweeps them aside in favour of unstated conclusions clandestinely arrived at by an unnamed intruder.

This is doing extraordinary violence to professional integrity and scientific procedures and must undermine its credibility and motives.

The IRM states that the NWDT in 1979 opined that without Narmada Sagar a 35-foot higher dam at Sardar Sarovar would compensate for the loss of irrigation and power benefits. The Nigam had also at some point suggested that storage could be created in the command area. "Either of these changes would alter in the most fundamental way the nature of the Projects. This, in turn, would change the environmental and social costs to an extent and in a manner which have not been assessed." This is a complete non sequitur. The NWDT's statement was only a hypothesis, not the actual project design under implementation. And the Nigam's idea of intermediate storages in the command area is recommended practice for fine-tuning water releases and water management in large commands and if and when adopted would be independently examined for its environmental impacts. This reference of the IRM is again a total red herring.

The notion that Narmada Sagar will not be built is a figment of imagination. In fact, the IRM itself showed knowledge of the fact that a construction contract for it was awarded on May 5, 1992.

Reservoir Simulation Studies and Canal Sizing

The IRM's other fear was that if Narmada Sagar was not built, there would be insufficient water to meet the designed irrigation and power projections at Sardar Sarovar. While the NSP might at worst be delayed by some years there is nothing to suggest that the complex, consisting of Indira Sagar (the mother dam), Maheshwar and Omkareshwar, will be abandoned. Even were that to happen, the Narmada Planning Group, the Operations Research Group, Baroda, and the Central Electricity Authority had carried out a series of reservoir simulation studies and effects at SSP in case of delayed upstream development. Single (SSP) and multiple (3-dam NSP complex) reservoir simulation studies with different hydro-power and canal sizings were carried out and it was established that even in the worst case scenario of no upstream development, SSP would perform.

The NWDT allocates 9 MAF out of the Narmada's total 75 per cent dependable flow to Gujarat, 0.5 MAF to Rajasthan (both regulated at SSP), 18.25 MAF to M.P. and 0.25 MAF to Maharashtra. Gujarat's own water use plan entails using its 9 MAF share of Narmada waters, 0.3 MAF available from rivers en route the SSP Canal, and 2.7 MAF from groundwater. Together with Rajasthan's allocation of 0.5 MAF, which has to be delivered at the Barmer border through the SSP Main Canal, Gujarat will be ensured 9.5 MAF even without upstream development with a 2.81 MAF carry over from the SSP reservoir which is within the design parameters, flows being diverted through the 40,000 cusec Main canal and network throughout the year as required. (NPG, November 1989) These studies were made available to the IRM and were simply ignored.

The IRM also wrongly assumes that the bulk of Gujarat's 9 MAF share must come from NSP. This is not so. Even today, the regulated releases from completed upstream Narmada Valley dams at Bargi, Barna, Tawa and so on average 0.4 to 0.5 MAF during each of the eight non-monsoon months. Taking a low average of 0.3 MAF, this will yield 2.40 MAF annually at SSP. Add to that live a storage of 4.43 MAF at SSP after four months of monsoon filling and a net irrigation of 2.86 MAF during the four monsoon months, and minus 0.35 MAF of post-monsoon evaporation. This gives a water budget of 9.34 MAF. Seventy five per cent dependability certainly implies low flows in three out of four years. But it equally implies excess flows in the remaining three years. The IRM is wrong again. Any remaining shortages in a succession of bad seasons could be made good by short-term mining of groundwater, adjustment of cropping patterns and water conservation measures (GOG, August 1992) with additional recharge in the ensuing good years within five-year, short-term hydrological cycles.

The IRM conjured up yet another issue. The SSP Main Canal has a capacity of 40,000 cusecs or 2.4 MAF per month at which rate the total live storage of 4.77 MAF could empty out in two months. It then went on to cite a World Bank consultant, Prof. Nathan Buras, as stating that the Canal would flow at less than half its capacity two-thirds of the time and the designed flow rate would be achieved only 5.6 per cent of the time. His conclusion was therefore that the Canal was probably larger than could be economically justified.

This early study was followed by a Bank letter of June 10, 1993 suggesting that the Main Canal should be enlarged to carry between 44,000 to 50,000 cusecs during critical periods of plant growth. The Bank pointed out that the flexibility and quality of the irrigation service required more than merely meeting normal peak demand. It should also be capable of meeting prolonged drought spells likely to occur between mid-August to mid-September which were also critical for plant growth. Canal capacity must be capable of meeting such emergencies. The Nigam studied the matter and decided that a 40,000 cusec capacity was sufficient as the Main Canal could be run carrying up to 45,000 cusecs over short spells when required by utilising the available freeboard. (GOG, August 1992) A larger canal would also enable the Nigam to utilise spill waters and divert these for augmenting water supplies in scarcity areas, for irrigated afforestation and for leaching saline lands. A smaller canal would fail at critical times when crop-life saving irrigation was most required. Going by its reasoning, the IRM would probably rate all lifeguards as idlers who do not merit their keep!

Upstream Impacts

The upstream impacts of SSP and NSP as of any other dam are manifold. Forests and crop lands are liable to be submerged. Erosion from degraded forest lands in the catchment could reduce the useful life of the dam through accelerated siltation whose delta-effect could entail backwater rise and inundation of additional areas. Submergence losses could add to pressures on lands and forests in resettlement or impact areas and displace wildlife, destroy habitats and ecological niches, and endanger or destroy valuable flora. This is well known by now and environmental impact studies and catchment area treatment and compensatory afforestation plans were mandated by the State governments and MOEF before forest and environmental clearances were granted. Work plans were required with deadlines for completion.

The SSP and NSP went before MOEF for clearance in 1983. Meanwhile, the Bank signed loan agreements which also stipulated its environmental concerns. In June 1987 MOEF granted conditional clearance for both SSP and NSP, followed by forest clearance. The Planning Commission's approval came in October 1988. The SSP was required to have a phased catchment area treatment (CAT) programme, plans for compensatory afforestation, command area development (CAD), and studies of flora and fauna, health aspects, seismicity, and the carrying capacity of the surrounding areas. The NVDA was required to undertake similar studies. Both were to complete these studies by 1989, keep MOEF informed and implement these programmes *pari passu* with progress of works on the projects.

The IRM found fundamental problems of compliance, with unconscionable delays and "institutional numbness at the Bank and in India to environmental matters". It, however, reported the concern of Bank missions over continuing delays. Against this, the NCA maintained that despite some slippage, all surveys and studies and implementation thereof were under way and would be completed *pari passu* with the Projects. MOEF was sceptical and on one interpretation its conditional clearance "was deemed to have lapsed", a proposition that the project authorities and the Ministry of Water Resources strongly contested. The MOEF, according to the IRM, wanted the clearances "renewed". Its conclusion, judgementally, was that there was "gross delinquency" in complying with the Bank and GOI's own norms and conditionalities.

Having said this, however, the IRM stated that the "1983 overview of ecological studies by the M.S. University Baroda, presented enough information for the development of proper assessment studies in the upstream area". But it found the follow up "disappointing". Likewise, it noted the fact that an independent committee had been set up by GOI in 1984 under Dr. M.L. Dewan to recommend soil conservation and afforestation measures in the SSP and NSP catchments based on which GOMP had prepared action plans. Relevant information for initial planning and implementation was on board well in time. The real complaint is about a spillover of certain further studies beyond MOEF's somewhat arbitrary 1989 deadline which was totally unrelated to its more meaningful *pari passu* conditionality which has been scrupulously adhered to. The MOEF's obsession with unrealistic deadlines for meeting unrealistic conditionalities was picked up by IRM to damn the projects for so-called environmental delinquency way ahead of any real damage on the ground that anyone could point to even three years after the 1989 deadline. Critics opposing the projects in pursuit of their own agendas were in full cry even earlier. The title of a book published by Alvares and Billorey, *India's Greatest Planned Environmental Disaster,* (1988) is typical. A section of MOEF with missionary zeal had also long concluded likewise and made strenuous efforts to obstruct if not block the projects on ideological and imagined technical grounds. In 1989 deadline was one of these and was nowhere justified. The IRM carried this fetish further to lay down iron laws for environmental clearance that have perhaps never been demanded or strictly fulfilled anywhere.

Catchment Area Treatment and Compensatory Afforestation

The Dewan Committee reported in 1985 on the state of the Narmada catchment (Dewan, August 1985). It found forest area submergence likely to be 4523 ha in Gujarat and 35,325 ha in M.P. under SSP and NSP together. Much of this was degraded and fast deteriorating due to increasing population and biotic pressure. The forest in Maharashtra was formerly under teak but was found to be badly damaged by tribals and now very inferior. Patches of good forest stands remained in M.P. and Gujarat, increasing eastwards with sal and bamboo. The NSP has a catchment of 61,600 sq km of which the larger part is cut off by Bargi and 14 other projects leaving a free-draining area of only 24,800 sq km. Likewise, of the SSP's total catchment of 88,000 sq km, 72,000 sq km is cut off by the NSP complex and five other projects on tributary streams leaving a free-draining area of no more than 16,000 sq km. Thus, from the angle of siltation, both the SSP and NSP catchments benefit from substantial cut-offs by other projects which operate as silt traps above them. Noting the erodible state of the remaining free-draining catchment and actual experience elsewhere, the Dewan Committee underlined the importance of catchment area treatment. The three states collectively proposed treating 5100 m ha or 58 per cent of the basin at a cost of Rs. 1,500 crores. This was considered too ambitious and a more realistic plan drawn up to treat 3,700 m ha or 42 per cent of the basin consisting of priority agricultural and forest areas at a cost of Rs. 1,332 crores, including land shaping of 400,000 ha. In view of resource constraints this was further pruned to cover just under 1.4 m ha or 15.8 per cent of the basin consisting of high priority watersheds over the next 10 years at a cost of Rs. 520 crores.

The proposed CAT plan included afforestation of 600,000 ha of degraded forest lands, bunding, terracing, land levelling and shaping of 750,000 ha, grassland development, nullah training and bunding, gully plugging and construction of silt detention structures. The committee estimated the sedimentation rate in the SSP catchment at 4.44 ha m per 100 sq km, and 5.74 ha m per 100 sq km in the Narmada Sagar catchment.

By December 1986, a third revised Action Plan for compensatory afforestation in denuded and degraded forest lands in the SSP and NSP catchments had been prepared by NVDA, earlier proposals having been rejected by MOEF. This envisaged compensatory afforestation in 90,000 ha in western M.P. in lieu of 44,320 ha of forest land going under the two projects. (Revised Action Plan, 1986) Similarly, for Gujarat, compensatory afforestation was planned and is under implementation in 4650 ha of non-forest land in lieu of the forest lands being submerged by SSP and upgradation of degraded forests in double that area (9300 ha) within the project impact area. By the end of the 1992 planting season catchment area treatment and impact area afforestation in Gujarat had reached 60 per cent of the target and compensatory afforestation in Kachchh, to the extent of 70 per cent. (GOG, August 1992) In M.P. about 34,900 ha of CAT and compensatory afforestation had been completed in the SSP and NSP commands. (Conservator of Forests, Khandwa, 1992) In Maharashtra too progress had been made. All CAT being done on the basis of prioritisation of watersheds by the All-India Soil and Land Use Survey Organisation, and compensatory afforestation is being closely monitored. (NCA, July 1992) This shows action not merely *pari passu* with but in advance of engineering works on the dams.

The World Bank conducted its own studies on environmental impacts relating to both the SSP and NSP. (World Bank, March 1989) This too reported on degraded forests and river beds being designated as "forest", much of which had been severely damaged or had long disappeared on account of illicit felling, overgrazing, shifting cultivation, clear-cutting and nistar forages for a growing population. Heavy encroachments had taken place. These "phantom forests" lacked any under-storey vegetation, had been subjected to fires, and suffered from severe sheet erosion in the NSP catchment. But some high quality forest would also be inundated. It also noted that the CAT and plantation programmes had taken into account fodder, grazing, fuelwood and nistar requirements.

The IRM, however, came up with other objections. The test for implementation, it argued, is whether or not there is adequate compensation for the forest lands being lost by submergence. "The straightforward 1 : 1 non-forest land or 2 : 1 forest land replacement is the real estate component of compensation. Other factors from botanical composition to long-term viability, are just as important, if not more so". In other words, quality is more important than quantity. In Gujarat, compensatory afforestation in Kachchh would be replacing "productive forest land" with "less productive", cheap, "agriculturally marginal" lands. In M.P. and Maharashtra much of the afforestation would be on degraded forest lands. Thus "the true ecological cost of forests submerged by SSP is not reflected by the compensatory afforestation efforts". What is being done is not doing any harm and "may even do some good", but the vital issue "is the principle that quality, not only the amount, of India's natural forest cover should not be allowed to be reduced further".

This is one of the IRM's no-win, heads-I-win, tails-you-lose arguments that is really quite ridiculous as it amounts to saying that no forests can be cut at all. Most of the forest lands being submerged are merely lands under the jurisdiction of forest departments (including river bed) and are not forested at all or are severely degraded or "phantom" forests. Repeated burning in some places has prevented regeneration and caused regression. Barring a few patches, the notion that all or most of the submergence area is under lush, pristine forests is an absolute myth which bears no relation to the ground reality. The extreme absurdity of the IRM line is seen in its lament over the diversion of the scrub Taloda forest in Maharashtra. "The high quality of the forest that originally existed in the area means that the lost opportunity in terms of potential forest development has been high".

The facts are otherwise. Vanishing forests are being restored to health. Both quantitative and qualitative aspects should distinctively improve. Other critics often sneer at compensatory afforestation as "plantations" or monocultures that do not replicate the original natural forest. This need not be so in degraded forest areas where the original root-stock survives and is capable of regeneration with some protection and densification. And the notion that a forest in Kachchh is a poor bargain has no merit. Reclaiming the desert is of immense value and preventing its spread and inducing favourable micro-climatic changes is also a considerable gain. Just as desertification has resulted in an exodus of people, re-forestation or afforestation will bring these areas back to life. And why should not forests in Kachchh be "productive" and how is their local, regional and global ecological value any less than any other forest elsewhere? Indeed, in larger ecological terms the trade-off of lost forests to regenerated or new forests in the SSP catchment and command is hugely positive by a factor of, maybe, 100 : 1.

The SSP will submerge 4523 ha of degraded forests bearing 981,000 trees. As against this, 11.6 m trees are being planted in Kachchh (of which 6 m had been planted by 1992). (SSNNL, 1992) 75 m trees are to be planted along 75,000 km of canals. Degraded forests with an average 0.1 per cent density are being upgraded by planting millions of trees. The dismissive attitude of the IRM to all this is breathtaking. It has no use for restorative ecology and, like many critics, is content with nostalgia and demagogy. And, having forecast the worst, it concludes with the observation: "The paradox we found is that in many areas, mitigation is actually ahead of the assessment"!

Siltation and Backwater Effect

The IRM and, taking off from it, other critics, have expressed grave apprehensions regarding the very considerable inundation and displacement likely to be caused by the backwater effect of delta formation at the upper edge of the SSP reservoir.

Fears of a drastic foreshortening of the life of the dam because of a far higher-than-assumed rate of sedimentation are unfounded. To begin with, the IRM takes no account of the fact that the assumed siltation rate in the project design of 5.34 ha m/100 sq km is less than the average rate of 5 ha m/100 sq km observed by the Dewan Committee in 1985. With CAT programmes, the actual rate is likely to fall. The construction of the NSP complex and dams on tributary streams between it and the SSP will act as silt traps and reduce the residual silt load flowing into the SSP reservoir much of which lies in a long, narrow, rocky gorge.

According to a study done by the Narmada Planning Group the life of the SSP reservoir ranged from 337 years (if the NSP is simultaneously completed) to 180 years (if the NSP is completed with a 25 year delay). A five year time lag in the NSP's completion would give the SSP a life of 222 years. However, CAT programmes will make a difference and even no more than an assumed 25 per cent efficiency factor would enhance the life of the dam by another 55 years. Construction of more dams on tributaries of the Narmada below the NSP will further improve the situation. Taking all factors into consideration, therefore, the life of the SSP can be expected to be between 333 and 437 years. (Raj, March 1991)

The backwater profile has also been carefully assessed. (CWC, 1992) The Sardar Sarovar Dam site has an estimated 1000-year design flood of 3.07 m cusecs which the spillway can safely pass. But it was agreed among the States concerned that property may only be acquired up to the level of the highest observed flood discharge of 2.45 m cusecs at Garudeshwar in 1970. Assuming a silt rate of 5.34 ha m/100 sq km per annum initially, falling to 3.23 ha m/100 sq km after 200 years, the CWC projected what might be the ultimate situation. The backwater effect was then calculated on the basis of a U.S. Army Corps of Engineers' formula for floods at the observed and 100-year peak levels without NSP. This indicated marginally higher backwater levels with reservoir siltation up to 94 km and a decline beyond that, with levels after 155 km actually lower than before sedimentation. With NSP in place, these levels would be still lower. Even the marginal effects anticipated will only manifest themselves 50 years after the commencement of dam operations.

In no project existing for the past 30-40 years has there been any complaint of backwater submergence along the reservoir fringe. On the contrary, the earlier practice in Gujarat of acquiring all lands up to the design HFL (highest flood level) was abandoned in 1964 when landholders represented that they were being needlessly deprived of valuable lands which were seldom if ever inundated. The new norm adopted was to acquire lands midway between the FRL (full reservoir level) and HFL. In 1990, following further representations, a strip six feet below FRL (which is again seldom flooded) was returned to landholders who were prepared to accept occasional short-term inundation. So, far from residents along the reservoir fringe fearing backwater submergence, they are prepared to live with very occasional flooding rather than lose these prime lands whose moisture regime gives them a special value. (GOG, August 1992) The extreme alarm sounded by the IRM and other critics is unfounded.

Flora and Fauna

Preliminary floral and faunal surveys of the SSP impact area were conducted in all the three states which provided basic data for planning. More detailed studies were envisaged and were likely to be concluded by 1992-93. None of the surveys in Gujarat revealed any rare or endemic species or any endangered species of plants or animals or aquatic fauna or fauna. In fact, deforestation and degradation of the forests and faulty land use practices have destroyed or depleted most plant and animal life except in pockets.

The situation on the left bank of the river is somewhat better in Gujarat which had accordingly extended its surveys to cover the adjoining Shoolpaneshwar sanctuary which is being sought to be upgraded and made into a sloth bear reserve. This proposition has, however, run into opposition as it might impinge on certain forest villages and affect their grazing and other rights. In the command area, plans are being prepared to establish and enrich the Nal Sarovar bird sanctuary, the Little Rann of Kachchh wild ass sanctuary and the Vedavelar black buck sanctuary. Thus, the end result is likely to witness an enhancement of flora and wildlife in Gujarat.

The position in M.P. is rather more mixed as degraded forests and depleted floral and faunal conditions in some areas are balanced by richer areas elsewhere, especially with regard to wildlife. While the creation of a number of sanctuaries has promoted wildlife in M.P., some wildlife in the submergence area might be affected unless migration corridors or other means of moving these animals can be devised. Protection of anjan-growing forests in the Khandwa region has been mentioned as a priority and creation of new sanctuaries to preserve game and plant life. As always, the right balance will have to be struck between this very laudable objective and the protection and enhancement of the lives and economy of the resident population which should not be crowded out. Wood from clear-felling the submergence area will cater to fuelwood needs for up to a decade but alternative sources of fodder, fuel and employment are necessary. Plans for these are included in the CAT and compensatory afforestation programmes but the propagation of improved chulahs, solar cookers and employment generation schemes must be vigorously pursued. Attention will also need to be paid, especially in Gujarat, to the interesting reptiles and amphibian species in what is a transition zone.

The region is already burdened far beyond its carrying capacity in terms of fuel and fodder. In the Gujarat sector, for instance, the carrying capacity on both sides of the SSP reservoir is estimated at 18,000 and 8,000 cattle units on the right and left bank respectively. As against that, the actual cattle population comes to between 50–70,000 and 26,000 respectively, excluding goat and sheep. Likewise, firewood consumption on the right and left bank is estimated at 40,000 and 7,200 tonnes per annum, whereas the productive yield is no more than 4,000 tonnes on the right bank and, more comfortably, 15–20,000 tonnes per annum on the left bank, presumably including the Shoolpaneshwar sanctuary. (Sabnis, March 1991) These problems are aggravated by a growing population and need to be addressed irrespective of the SSP. While the SSP will certainly impact on this situation it has also afforded opportunity and funding to undertake ameliorative action for the first time.

Cultural Impacts

Surveys by the M.S. University, Baroda, show that the tribal people are unhappy with their lot and wish to improve their lives through education and wider employment opportunities. A number of recommendations to this end have been made. (M.S.

University, October 1992)

The Narmada is a sacred river and attracts tens of thousands of devotees who worship at various shrines which are the venue of important fairs and festivals and circumambulate the river. The parikrama pilgrimage is especially auspicious. The SSP and NSP projects have carefully avoided submergence of any protected monuments, especially those at Maheshwar and Omkareshwar. But some lesser shrines will be affected and are being relocated at higher elevations like Hamfeshwar in Gujarat, or the deity is being ceremonially moved, as in the case of the Shoolpaneshwar Temple at Manibeli, whose spirit was ritually invoked to relocate in a new temple at Garudeshwar. Some historic sites are also being relocated. The parikrama route is being redrawn as portions of the existing route will undergo submergence. No problems are envisaged and archaeological teams are surveying the area to conserve pre-historic relics of the Narmada Valley, home to an ancient civilisation.

With the creation of lakes behind the dams, shoreline vegetation and aquatic and animal life is expected to improve. Sanctuaries are being created and tourism will receive a fillip.

Dam Safety and Seismicity

Fears relating to the safety of the dam or the neighbourhood from earthquakes or reservoir-induced seismicity can be discounted. The strongest earthquake ever experienced in the region registered M 5.6 at Bharuch in 1970. The geology of the dam sites and seismic coefficients were worked out after great deliberation and wide consultation with Indian and international experts. The NVDA prepared detailed position papers for the NSP complex and conservative designs have been adopted despite higher cost as a matter of abundant caution. The World Bank reviewed these designs through independent experts and there is no reason to suppose that the safety of these dams is imperilled notwithstanding the M 6.1 earthquake that struck Killari-Latur 300 km away in Maharashtra on September 30, 1993. Both the SSP and NSP sites are being instrumented to measure seismic events and the effects of reservoir loading which could trigger reservoir-induced seismicity of a modest order in the initial period. RIS cannot in any case be of a magnitude greater than natural seismic events but can operate as a triggering device. Therefore, the provision made for the natural hazard will in all cases contain the smaller element of risk inherent in RIS.

Downstream Effects

The Narmada meets the sea 180 km below the Sardar Sarovar Dam. Impoundment will naturally modify flow and silt regimes, affect river and coastal morphology, salinity, groundwater and fisheries and have other effects. The IRM was highly critical of the lack of environmental impact assessments of downstream effects apart from the preliminary analysis made by M.S. University in 1983. The IRM was informed of flood control benefits, assured that any increased river bed scour below the dam would be manageable and that salinity would not be affected.

As in other sectors, the IRM was not satisfied and engaged an unnamed independent expert. It collected or collated other information from a variety of sources and made some geomorphological studies pertaining to the Lower Narmada and its estuary "although it was never in our terms of reference". As a result of these sundry efforts it concluded that there was room for concern about salinity, pollution and water quality, fisheries and the impact of subsequent urban, industrial and agricultural developments in this region. It was also sceptical about flood control benefits since the SSP has limited flood storage and is not intended to provide a flood cushion as a primary operational objective. On what is this pessimism founded?

Flood Mitigation

Recurrent floods in the lower Narmada endanger Bharuch city and about 200 villages with an aggregate population of over 750,000. As many as 325 lives were lost and livestock perished and huge crop and property damage was sustained in the 1970 floods. With the SSP in place these areas will be protected. There is a flood cushion of five feet between the FRL (full reservoir level) and maximum water level of the SSP at 460 feet which will be used to moderate floods and facilitate flood routing with the aid of a hydro-met network that is being established. This will be a major boon. The IRM is quite wrong in its estimation of flood hazards.

Low Salinity Risk

Storage and diversion may normally be expected to reduce downstream flows below a dam. However, for a variety of reasons this will not happen in respect of the SSP for 35 or more years. This is because, until the SSP and Rajasthan commands are fully developed to utilise the designed canal flows, storage releases will continue to pass through the river-bed power house even during the lean months, actually augmenting present low flows by a significant order. Further, the NSP complex and other Narmada Valley projects in M.P. will also take time to complete and achieve full utilisation of that State's share of 18.25 MAF of water until when the surplus will pass below SSP.

SSP irrigation development is envisaged in three stages and until Stage-III, commencing 35 years after the start of reservoir operations, the dry season flow, presently down to 1000 cusecs, will be between 15–20,000 cusecs. Even now, with regulated releases from the Bargi reservoir, the salinity wedge has been pushed further down the estuary. Silt entrapment behind the SSP and upstream reservoirs will deepen the downstream channel, improving conditions for fisheries and navigation. Bank erosion will be monitored and protective measures initiated such as mangrove development in Aliabet island. Agricultural developments, groundwater recharge and industrial and municipal uses will not be affected.

This being established, and the near and medium-term future being assured, there is an air of unreality in wanting here-and-now studies, which of necessity would be speculative, of impacts 40 or more years down the line by when other factors would have come into play.

Among these is the possibility, already envisaged, of diverting up to 2.09 MAF of water from the west-flowing rivers of south Gujarat northwards from the Damanganga river to the Tapi (Ukai dam) and thence to the Narmada below the SSP, as well as from the National Water Grid. Both these have been looked at by the National Water Development Agency, which is engaged on national water grid studies, and hold out promise. It is estimated that 1.63 MAF could realistically be diverted to the Narmada from the South Gujarat rivers within the next 40 years at a cost of around Rs. 1,000 crores. What emerges is the low utility of insisting on studies today of situations likely to obtain 30-40 years hence. To that extent the IRM was being unduly alarmist without reason. (NPG, November 1989)

Rich Fisheries Prospect

The IRM was concerned that with changes in the river regime after the SSP is completed, changes in water quality from reduced flows could effect migration and spawning of fish and the breeding of hilsa, prawn and mullets, prize Narmada catch. The Narmada is a major source of hilsa spawn seed for stocking and any adverse impacts would have far-reaching consequences. Changes in flow patterns and the nutrient load and estuarine and coastal influences could have cumulative effects, some of which might be adverse. The Nigam's views to the contrary on some of these points notwithstanding, the IRM complained that it was being given assurances rather than assessments and several doubts remained.

The M.S. University had looked at Narmada fisheries in its 1983 quick survey. It established that the river and estuary provided half of Gujarat's inland fish catch, thus constituting an important asset in terms of food, income, employment and seed stock. The study saw rich fishing possibilities in the SSP reservoir with a conservatively estimated yield of 80 kg/ha year or a catch of 96 tonnes in the small Gujarat portion of the lake alone, the largest gains going to M.P. Downstream hilsa fisheries could be affected in Stage-III of the project after 35 years but a second hilsa migration run during winter might be sustained if a short duration flood release from the dam could be ensured. Breeding and rearing of hilsa in fresh water had also proved encouraging. The M.S. University concluded that projections made over the next 85 years indicated that losses on account of hilsa and prawn would be more than made good by gains in other varieties of inland fish. (M.S. University, July 1983)

Hilsa has been found to migrate only 106 km up river and so its movement will not be impeded by the SSP dam. Raising of hilsa in the Ukai reservoir has after some initial difficulties seen production increase from 863 tonnes in 1984 to over 2,000 tonnes in 1987. If this can be repeated in the SSP lake, the prospects would appear very bright.

The Mahaseer is a migratory fish in the upper Narmada but

has been successfully established by now in several reservoirs. As with the SSP, the NSP complex of lakes will replace the riverine fisheries at far higher levels of productivity, given proper stocking, management, training and marketing facilities. In 20 years, reservoir fisheries in the NSP and SSP complex are expected to yield an annual catch of a little under 4,000 tonnes as against an inland commercial fish production in all of M.P. of around 400 tonnes. (Levenhagen, March 1989)

To this must be added the possibility of developing extensive fisheries in ponds and other water bodies in the command area and in the larger canals. Fishery possibilities will be enhanced in the river and reservoir. Overall, the dams should stimulate fishery development quite considerably.

Techniques for artificial fertilisation of hilsa and freshwater prawn have been developed and plans have been prepared to establish hatcheries and achieve an annual stocking target of 10 m hilsa fry and post-larve giant freshwater prawns in the Narmada basin. Most of the 8288 active fishing families in 453 fishing villages along the entire length of the Narmada could be reemployed in these new fisheries. (Central Inland Capture Fisheries Research Institute, 1991)

Overall Downstream Outcomes

The downstream environmental impacts have been noted and reviewed and could perhaps have been further mitigated had the NWDT acceded to Gujarat's request for a 0.7 MAF allocation for downstream (ecological) releases. The NWDT did not go along with this argument presumably because measurable impacts from reduced flows will only be manifest after Stage-II of irrigation, 30 years down the line. Meanwhile, apart from additional flows being released through the river-bed power house turbines, the SSP and other dams are expected to spill in good years, making available an unallocated source of water which Gujarat has been permitted to use with due notice to upstream states.

Assurance about the downstream environment also comes from a consultancy study. This looked into the whole gamut of possible negative impacts and notes that "there are no downstream impacts whose magnitude and effect are such as to cause doubts to be cast over the wisdom of proceeding with the SSP, provided that appropriate monitoring and mitigation measures are applied". These include deliberate flushing of the river to help fish migration and spawning and mitigate adverse health factors, and the restoration of previously degraded forests in Aliabet and other islands. It also advocates the option of guaranteeing a minimum downstream flow in the lower Narmada (in due course) for which the NWDT had declined to make a separate allocation. This might entail less irrigation supplies but would enhance power generation, improve water quality and pollution dispersal in a fast industrializing zone, and benefit health. The matter is worthy of consideration. But nothing hangs on a here-and-now decision. (Wallingford, February, 1993)

Health Implications

Existence of waterbodies and introduction of irrigation do create situations favourable for the spread of waterborne diseases. Health surveillance over river valley projects is therefore of great importance. This factor was not ignored in the Narmada Valley. The M.S. University conducted a health survey in 1983 in the Gujarat area and found scabies and skin diseases ranking third, both being major scourges in the tribal belt.

The World Bank rated malaria, schistosomiasis and filaria as the three greatest hazards, but subsequent surveys by the National Institute of Communicable Diseases and, separately by the Bank and WHO established that there is no risk of schistosomiasis in the Narmada Valley. Filaria was found to be prevalent near the SSP dam site while malaria is more widely present but at a low level. The IRM, however, quoted a report by N.L. Kalra (February, 1992) criticising aspects of health planning under the SSP. In particular, he cited the so-called dyke-villages by the dam as a "death trap for malaria" and said this was liable to lead to Japanese encephalitis. Canal and drainage works with borrowing of earth from deepened village tanks amounted to "taking malaria to the doorsteps of villagers". Taken with reports of some malaria deaths near the dam site, this somewhat alarmist finding, partly torn out of context, was used by the IRM to paint a picture of something approaching a medical catastrophe.

Closer investigation showed that after continuous years of drought, heavy rains in Gujarat in 1988 and 1989 had resulted in a higher incidence of malaria in most of Gujarat. This puts Dr. Kalra's findings in perspective. It was also not reported by him that the incidence of malaria in the 19 Gujarat SSP villages as also in the dyke villages had come down in 1990 and 1991. The malarial parasite had in some cases been brought by outside labour contracted to work at the dam site.

On the other hand, availability of water throughout the command area will improve personal hygiene and sanitation and bring down the incidence of water-related diseases.

Drinking Water Boon

More than irrigation, invaluable though that is, maybe the greatest single boon of the SSP will be the provision of assured, potable water to 40 million people (by 2020) in 135 urban and 8215 villages in North Gujarat, Saurashtra and Kachchh. (Some of them currently uninhabited.) All village ponds and tanks will be filled by the canals so as to ensure sustained year-round supplies irrespective of rain or drought. Today, millions face acute water shortages every year and are compelled to migrate with their livestock in bad years which recur with regular frequency. Even otherwise, towns and villages are ferried water by train and road tankers on an emergency basis at enormous cost for minimal supplies. The havoc this plays on health, leisure, production and productivity is incalculable. Women, the girl-child included, spend the best part of their daily lives, throughout their working life, walking endlessly, day after day, fetching water and foraging for fuel and fodder. This hopeless drudgery is at the cost of their childhood, health, education, creativity and leisure. The human cost enters no calculus of cost-benefit analyses that critics or even governments project. It is a human tragedy of the first order.

The situation is in fact getting worse with the steady lowering of the groundwater table on account of excessive withdrawals, some tubewells going down to 300 m. There has been salinity ingress in coastal areas up to 6 km, while water quality has been affected in a number of areas with nitrate and flouride contamination. During the 1985-87 drought, Gujarat spent over Rs. 1500 crores on supplying drinking water and other relief measures. The cost of supplying water to Rajkot city came to Rs. 250 per cubic metre or double the cost of desalination. As many as 1.6 m cattle with their herdsmen migrated over distances of 400–600 km.'

It is against this background that the NWDT enjoined that 1.06 MAF out of the 9 MAF allocated to Gujarat be devoted for domestic, municipal and industrial uses, the allocation specifically for drinking water being about 0.853 MAF. The Gujarat Water Supply and Sewerage Board (GWSSB) and the State's Industries Department accordingly worked out plans for utilising this water. Offtake points have been identified all along the canal system. Plans have been formulated over a 30 year perspective up to 2021. Allocations in terms of litres per capita/day have been fixed at 140 and 100 for towns with and without drainage facilities and 70 for rural areas. All villages in Saurashtra and Kachchh and all problem and no-source villages in North Gujarat are to be supplied. Domestic requirements within the SSP command up to the Mahi river will be met from groundwater sources with canal supplies and local sources serving the areas beyond. It is by such adjustments that the initial proposal to supply 131 urban areas and 4719 villages was subsequently upgraded to supply a larger number of communities.

This increase in numbers and the absence of any cost figures for the drinking water programme in the SSP budget has been used by the NBA and other critics to debunk the scheme as a hoax. The SSP will only guarantee the water and carry it to agreed outlets from where the GWSSB will take over, carrying the supplies by existing or new pipelines or water channels to consumption centres. This will be a GWSSB project and will be separately funded. There is nothing unusual about this and it is standard practice. As for the number of villages and towns to be supplied, the total quantum of water involved is so modest in relation to irrigation demands that a small adjustment can always cater to large additional populations.

The notion that drinking water supplies are a cosmetic add-on or afterthought to brighten the image of the SSP is fanciful. Reaching water to Saurashtra and Kachchh was embodied in Gujarat's thinking from the very early days and was strongly argued before the NWDT. This was a planned decision and no latter-day image-building supplement.

Command Area Concerns

A number of environmental concerns have been raised by critics regarding the command area. These range from lack of water to soil and irrigability factors, risks of extensive waterlogging and salinity, cropping patterns and yields, water management and water-use efficiency, and the alleged lack of studies pertaining to these matters.

The IRM is sweeping in its criticism. It places its remarks in the context of the World Bank's "India Irrigation Sector Review" published in 1991 which pointed to a number of shortfalls in the design and operation of Indian irrigation schemes and made recommendations for improved water management. It also advocated avoiding new starts before completing on-going projects to which it traced many ills of construction, delays, poor operation and maintenance, resulting in poor deliveries and consequent inability to price water at its true economic and service cost. These factors have been highlighted by Indian commentators and the point is well taken. The design and planning of SSP takes all this into account and in that sense represents a leap forward in the conceptualisation, engineering and operational management of irrigation. This is what one would expect and applaud.

On the contrary, the IRM, like some others, is sceptical and predicts failure because this was not the way things were done or how things performed previously. It admits of little possibility of learning from experience and, while urging improvement, insists that improvement is impossible. Once again we have a no-win argument. Where the Bank appraisal teams found an exciting new departure in the SSP, the IRM sees none. And again, much is made of some initially-prescribed target dates for "work plans", first 1985 and then 1989, not being met. Yet much work is in hand and operational findings are being fed into the system well ahead of project schedules so as to ensure their implementation. Here too, *pari passu* action in relation to a huge command area is unacceptable to the IRM and the ultimate is sought to be programmed with the preliminary. This is quite unrealistic and it would be interesting to know where in the world such a degree of complete and detailed advance action has been laid down and accomplished? Theory or arbitrary deadlines apart, the IRM and others have failed to show how these so-called delays have been detrimental to project planning or execution in practice. The SSP carries the usual hazards of complex planning and innovation; no more, no less.

The fact is that the SSP sets out to develop a most advanced irrigation system with a number of built-in checks and balances which are subject to fine-tuning on the basis of close monitoring.

Irrigability, Waterlogging, Salinity

The gross command area has been divided into 13 regions on the basis of agro-climatic and socio-economic features within which irrigability has been determined under six soil classifications including drainage capability. Groundwater profiles, existing and as likely to develop with irrigation, have been studied and for the first time in India, conjunctive use of surface and groundwater has been built into the irrigation design from the very start. Within these parameters, a feasible cropping pattern has been prescribed for each region and category with a regional allocation of water, linked to groundwater use to ensure vertical and, where necessary, surface drainage. Water availability for the command is placed at 12 MAF, with 9 MAF from the SSP reservoir, 0.3 MAF from en route rivers and 2.7 MAF from utilisable groundwater within the command. (NPG, November 1989; Dave, K.M., 1991)

Waterlogging and salinity are sought to be minimised by lining the entire canal system from the Main and branch canals down to the lower level network up to 8 ha outlets in order to reduce seepage losses. As extensive rather than intensive irrigation is planned, the water delta is being limited to 53 cm against a normal delta of 75 cm in many existing projects. Command area development will ensure construction of field channels and field drains and land shaping and consolidation where necessary. Strict water control is sought to be exercised through application allowances related to regions and irrigability classifications initially through semi-automatic and later through fully automated canal operations. Volumetric supplies will be made through rotational releases (warabandi) and conjunctive use of water. Active involvement of farmers through participative management and organisation of water user associations is expected to impose a self-regulatory regime based on the principles of an irrigation democracy. Burrow areas in the command have been kept to a minimum by using earth exacavated in the process of de-silting and deepening village ponds and drains. This should help reduce seepage losses.

A network of 87 piezometers and 18 observation wells have been in place up to the Mahi for some time and another 144 piezometers and 33 observation wells are now ready in the areas beyond. Regular observations, together with groundwater aquifer modelling studies, will enable project authorities to monitor groundwater levels. As soon as rising water tables approach critical levels, project management authorities will drill and operate tubewells and induce farmers to do so likewise by fine-tuning water allocations. Likewise, where it is so indicated, saline water will be pumped out and disposed of through surface drains. About 0.66 MAF of unusable groundwater is to be removed in this fashion. The probable quantum of waterlogging in the command under a worst case scenario is likely to be no more than 5 to 7 per cent, given proper water management. With vertical and surface drainage this can be brought down further. Surface drainage will be required over 65 per cent of the command and in large depressions and burrow areas. In several places the quality of groundwater will distinctly improve. (NPG, April 1993)

Doubts have been expressed as to whether these paper refinements will actually work. An experienced irrigation economist has certified the plans as workable after taking a close, hard look. Apart from project pumping for groundwater use, he is satisfied that requisite private investments in dugwells and tubewells will be forthcoming in the SSP command. In view of the modest water allowance provided, farmers will have strong compulsion to make good any crop-water deficiencies. The development of a good groundwater market in Gujarat should further facilitate requisite investments in private groundwater development. The danger he sees is in farmers abandoning the prescribed cropping patterns and adopting crops like paddy and sugarcane which are water guzzlers. (Dhawan, B.D., September 1990)

This view is further reinforced by another expert opinion. Given sound management, "there is no reason for waterlogging and salinity to present major problems, although in some limited areas economic considerations may make it desirable to restrict the extent of the command". Irrigation was introduced in the adjacent Mahi Right Bank Command in 1980 and "severe water-logging" developed in the absence of proper drainage. A drainage master plan covering 44,000 ha was thereupon drawn up and with improved drainage and by halving the water allowance "the situation appears to have to a large extent been remedied". (Wallingford, March 1993) Waterlogging in the Ukai-Kakrapar and Kadana commands in Gujarat is in the range of 2.5 to 3 per cent. There are 14 sugar mills in the region, all of them outside the SSP command. Two factories are proposed to be established within the command but they have been told that the requisite water supply will not be assured and the total area under cane in the command will not be permitted to rise above 0.5 per cent as planned.

There are certain problem areas, with impervious vertisols and inceptisols which swell with water, where irrigation, unless carefully planned, could have adverse results. Among these are the Bhal and Bara tracts along the Gulf of Khambat in Saurashtra. Management strategies here include permitting only low water intensity crops, simultaneous excavation of drains, avoiding direct flow irrigation and instead providing deep-cut water channels from which water will have to be lifted to irrigate the fields, thus ensuring limited application. Surplus monsoon waters will be used to recharge groundwater and, where necessary, vertical drainage will be provided.

Both as a water-saving device in the more arid areas of North Gujarat, Saurashtra and Kachchh in particular, and also as a means of averting waterlogging and salinity effects, thought is being given to encouraging sprinkler and drip irrigation. There are some experimental farms in Saurashtra where drip irrigation of cotton and vegetables has been highly successful, with nutrient doses also being fed through the system. The advantages are highly controlled irrigation and fertilisation, avoidance of land levelling, and savings in tillage. Water use can be brought down by 70 per cent with better yields. Construction of a micro-irrigation network can also be avoided with related land acquisition for field channels and drains. As against this, groundwater recharge will be forgone and there will be an energy component. The water saved could be used to extend irrigation to new areas.

No firm decisions have been taken, but the possibilities are there, especially for high value crops. Some very tentative but favourable estimates of costs and savings are available. (Raj, P.A., 1989)

Water-use Efficiency

In the light of all these provisions, the IRM's conclusion that water supply appears to be over-estimated and needs under-estimated is unwarranted. That hypothesis rests on the fact that planning on the basis of average rainfall ignores variability in precipitation and actual crops needs while an assumed 60 per cent water-use efficiency is much too high. True, rainfall may be deficient when required. It is precisely for this reason that canal capacities have been provided to ensure emergency supplies, supplemented by additional groundwater withdrawals to be made good by subsequent recharge. Similarly, a 60 per cent water-use efficiency is the national goal as against a 40 per cent average at present, with some systems dipping below that level. The more carefully designed SSP irrigation system is capable of achieving 60 per cent efficiency with good water management. Gujarat has been winning prizes for water-use efficiency in national competitions over several years. The targets have therefore been fixed on the basis of already demonstrated high-performance. Efficiencies of 66 to 71 per cent were achieved in the Aji-II, Demi-II and Uben projects, albeit in small commands, during 1990 and 1991 and is likely to cross 60 per cent against the 53 per cent already achieved in the larger Dantiwada project once on-going lining works are complete. (NPG, April 1993)

Equity for Saurashtra and Kachchh

The NBA and other Indian critics have sometimes argued that the SSP makes exaggerated claims to being Gujarat's "lifeline" as it will not impact on drought as much as made out and that not much water may be available or allowed to reach tail-end Saurashtra and Kachchh. The drinking water programme is also seen as so much window dressing or a pipe dream. More recently, great play has been made on the seeming injustice done to Saurashtra and Kachchh which have been notionally described as beneficiaries but will actually only receive marginal gains. This is a wholly uninformed view and ignores Gujarat's insistent pleadings for a larger water allocation and a higher dam precisely in order to reach the maximum possible benefits to Saurashtra and Kachchh.

Nearly 1.8 m ha of land spread over 3393 villages in 62 talukas in 12 districts will come under irrigation in Gujarat. About 75 per cent of this area is drought prone. Gujarat is not a water-rich state. Its utilisable water resources (including 9 MAF from the SSP) is no more than 21.46 MAF of which Saurashtra with its 71 rivers contributes 1.686 MAF and Kachchh with its 97 rivers no more than 0.326 MAF. The groundwater position in these two regions is no better, with few freshwater aquifers. The coastal areas and tracts adjoining the Great and Little Ranns are saline, with groundwater extraction leading to further saline ingress. This has affected water supply in 779 coastal villages. In addition, another 1207 scarcity villages face fluoride contamination of their water supplies. Rains are low to scanty over most of Saurashtra and Kachchh, with variability increasing sharply over large problem areas. Current water supply is extremely low, being 15 litres per capita/day in 30 per cent of the villages, 25 lpcd in 25 per cent of the villages and no more than 35 lpcd in the remaining villages. Urban supplies are also low and local resources can at best cater to no more than a third of the requirements of Saurashtra and Kachchh. The balance must come from outside and the only large source available is the Narmada. (NPG, April 1993; NCA, March 1992)

Saurashtra has a cultivable area of 4.2 m ha and Kachchh 2.37 m ha. At present, only 2,30,000 ha and 70,000 ha are irrigated in these two regions and with the completion of schemes under construction or contemplated, the total area that could ultimately be irrigated might reach 355,000 ha and 120,000 ha respectively. As against this, the Saurashtra Branch of SSP will irrigate 386,000 ha in Saurashtra and the Kachchh Branch 57,300 ha in Kachchh. All the 4877 villages and 90 urban centres in Saurashtra and 948 villages and 10 urban centres in Kachchh will receive drinking water from the Narmada canals, as well as all problem and no-source villages in North Gujarat. Both the Saurashtra and Kachchh branches will entail a number of lifts which will require energy, and falls on the other side from which some power will be generated. Studies are under way to see if the irrigation benefits of the Kachchh Branch can be extended to cover an additional

100,000 ha if additional water is made available. Savings through drip and sprinkler irrigation, utilisation of spills at Sardar Sarovar during the initial years and the northward diversion of waters from certain west-flowing, South Gujarat rivers are among the potential sources. Locations for on-line intermediate balancing reservoirs have been identified. The Saurashtra and Kachchh branches will also replenish all tanks and village ponds en route so that these become perennial sources.

The creation and enhancement of wild life sanctuaries in Saurashtra and Kachchh (black buck at Velavadar, wild ass in the Rann and the Great Indian Bustard in Kachchh), and the bringing of 5738 ha of land under compensatory afforestation in Kachchh together with canalside and roadside plantations in both regions represents a tremendous benefit. The greatest gift of the SSP will be drinking water, and here the coverage in Saurashtra and Kachchh is maximum. There will be a dramatic impact on migration of human and livestock populations, on drought-proofing, employment and incomes and on the quality of life, especially of women and children. To those who say that the promise of drinking water is a chimera, the answer is to invite them to see the transformation the Rajasthan Canal has effected in this regard in Bikaner, Jaisalmer, Jodhpur and Barmer where too the promise of drinking water was scoffed at for years. No more. Wallingford, the World Bank's consultant, has no doubt that the general quality of life including health will be enhanced in Saurashtra and Kachchh.

Those who talk of multiplying small water harvesting structures in Saurashtra and Kachchh ignore that fact that these are being constructed wherever possible. But what is simply not understood is that in dry years and critical periods, these small bandhs, wells and tanks run dry. The migrations that follow are recorded in census reports. A study was made of 25 small and medium dams in Saurashtra, all but four of them operating for over 10 years. Many of them were found to fill below half their capacity in 25 to 40 per cent of the years surveyed and indicated poor irrigation reliability. (Parikh, Vora, Alagh, June 1993)

Agricultural Dividends

The agricultural outcomes of the SSP are likely to be transforming in terms both of output and employment and, as important, agricultural stability, especially in the semi-arid and arid regions of the command which suffer scanty and uncertain rainfall. The project will lay the foundation for the next phase of Gujarat's poverty alleviation programme, the majority of cultivators in the command being marginal and small farmers and other weaker sections, including nomadic populations. For every tribal displaced in the catchment, seven will be benefited in the command while all PAPs will experience a quantum jump in their living standards within a few years. Wage levels are expected to rise with higher cropping intensities and all the related activities this generates. While the SSP will go far towards realising Gujarat's irrigation potential, the possibility of utilising any additional downstream supplies spilling over the dam is being kept in mind. These waters could be diverted through a low level canal to cater to the irrigation needs of a significant part of the command up to the Mahi, thus releasing the water saved in the Main Canal for additional irrigation beyond the Mahi. Depressions beside the dam have been dyked and will store 0.02 MAF which will help meet peak demands to some extent.

Cropping patterns will change in favour of superior crops and cash crops such as cotton and oilseeds. The sugarcane acreage is, however, being strictly regulated. Assured irrigation will promote use of high-yielding varieties. The increased value of crop output is estimated at over Rs. 900 crores per annum. Fisheries and dairying benefits would be additional. Critics have challenged these estimates as being overly optimistic. This is not so as they are based on yield levels already attained in comparable irrigated conditions in Gujarat. (NPG, November 1989)

Hydro-electric Power

Though the SSP is principally weighted in favour of irrigation, it will also augment power supply. There will be a river-bed power house (RBPH) with six reversible turbines of 200 MW each (1200 MW), capable of operating in the pump-storage mode with the Garudeshwar weir providing a balancing reservoir. In addition, there will be a canal-head power house (CHPH) with five 50 MW units. Total installed capacity at SSP will be 1450 MW excluding some limited generation at canal drops. The RBPH will have a firm or minimum year-round output of 415 MW and the CHPH a firm output of 50 MW. Actual energy production will be higher because of the larger availability of water that can be passed through the turbines during certain periods. However, as the irrigation and canal system develops and more water is diverted, continuous generation at the RBPH will tend to drop through succeeding phases of Stage I and II spread over 35 years till it falls to zero after 45 years when the entire canal system is fully operational. On the other hand, the CHPH will start with a continuous (firm) output of 24 MW, rising to 50 MW when the canal system is fully developed in 45 years.

The Western grid which covers Gujarat was short of 1,200 MW of peaking power during the Seventh Plan, the energy deficit being of the order of 330 m units in March 1990. Setting up additional generation capacity to meet peak loads, typically over six hours a day, is prohibitively expensive in view of the large idle time in between peaks and the impossibility of varying thermal output economically to any marked degree (except if gas-fired). Hydel power can, however, be augmented or switched off practically at will by regulating water releases through the turbines. It thus has a flexibility that makes it invaluable for meeting peak demands. The CHPH at SSP will be capable of operating as a peaking station, running at its full capacity of 250 MW for 163 days in the year (related to canal operations for maintaining full irrigation supplies). The RBPH will theoretically wind down to zero in 45 years provided M.P. utilises its entire share of 18.25 MAF within that period. Thereafter, any water savings, together with spills in good years and during the monsoon months, will enable it to generate power. Over and beyond this, the RBPH reversible turbines will be able to pump back water into the SSP lake from the Garudeshwar balancing reservoir during off-peak hours, using surplus base load power available within the Western grid, and recycle this same water to generate six hours of peaking power when required. Thus, the RBPH, far from becoming derelict within 45 years or only partly operational even earlier with increasing canal diversions, will function as a peaking power station with pumped storage throughout its life. (NCA, July 1990) Misconceptions on this score are widespread and the NBA and its friends have mistakenly written off the SSP power facility as a rapidly depleting asset and something of a white elephant. On the contrary, it will make a most useful and long-lasting contribution towards stabilising the Western regional electricity grid, allowing thermal stations maintenance downtime during the monsoon when it will be in a position to work continously at full capacity. This will enhance the thermal plant load factor and improve system efficiency and profitability. The unit cost of generation from the SSP hydel stations and its cost per unit of installed capacity will be lower than that of comparable thermal stations, especially if its secondary power generation is included. NSP power benefits are additional. Hydel power is also absolutely clean.

Navigation Possibilities

The Narmada is a long river but only the lowest reach, including the estuary, has been used for very limited navigation around Bharuch. Of course, country craft ply short stretches and ferry passengers and goods across the river. But the Narmada has never been a transport artery. The harnessing of the river will, however, create conditions favourable for inland water transport. Regulated releases during the lean season will improve drafts while pondages from the Sardar Sarovar dam up to the end of the ISP lake in the NSP complex will submerge falls and rapids to create a navigable fairway.

This thought imbued early planning and soon after the NWDT gave its award, the Government of India thought fit to have a feasibility study made of navigating the Narmada from Hoshangabad (637 km from the sea) or even from Jabalpur (a further 309 km upstream). A joint study was made with Dutch collaboration and a report submitted in due course. (CES and Haskoning, April 1985) The consultants were also required to examine the feasibility of navigation in the Main Canal, at least from the Sardar Sarovar Dam to a point on the Sabarmati river near Ahmedabad (230 km). Although IWT on the canal was found to be technically feasible and cost-effective, navigation was not recommended because of its impact on canal structures and planned irrigation operations.

The consultants found river navigation feasible and attractive with 20 m tonnes of coal on offer for movement from the Central

India, Western India, Pench-Kanhan and Singrauli coalfields to the west coast, rising to 30 m tonnes by 2010–2015. Upstream cargoes of petroleum products, salt, vegetable oils, pulses, fertilisers and unitised cargoes of up to 1.5 m tonnes were envisaged by 2010. The main poser was how best to pass the Indira Sagar, Omkareshwar, Maheshwar and Sardar Sarovar dams. Various alternatives such as single or twin locks, ship lifts and water slope systems were considered and appropriate recommendations made for each dam. Only in the case of Sardar Sarovar was a cargo transport system envisaged. Provision has been made in the layout of the dams and appurtenant works for the inclusion of navigational infrastructure at a later date as no firm decision on developing navigation has yet been taken.

Two types of barge fleets were considered, both requiring a draft of 3 m when loaded. The second type of river coaster was found more suited to estuarine and coastal transport from Bharuch to Ahmedabad-on-Sabarmati and ports along the Saurashtra, Kachchh and South Gujarat coast, each carrying 2000-tonnes apiece. Based on an annual operation of 300 days and a throughput of 20 m tonnes, fleet requirements were worked out with provision for moving 30 m tonnes annually in a post-2010 phase and even larger volumes thereafter with further investments.

The navigation option remains open and can be taken up after the existing project is complete provided no new developments are allowed to pre-empt it. The existing railway system and national highway run east-west through the Narmada valley but dip south after Khandwa, leaving a large part of the lower Narmada valley unserviced. This has retarded the development of this hilly and inaccessible tribal belt. A navigation link would be cost-effective and environmentally-friendly and would do much to boost the riverine economy. The notion of seeing barge-trains plying the river up to Hoshangabad or Jabalpur is not far-fetched. They could become major inland ports.

Upstream Narmada Developments

The Narmada Valley Development programme envisages the construction of a single terminal dam, the SSP, and the Garudeshwar weir just below it, in Gujarat. Another 29 major dams, 135 medium dams and over 3000 minor projects are planned in Madhya Pradesh in the middle and upper reaches of the Valley along the main stem of the river and its 41 tributary streams to irrigate 2.755 m ha of land and generate 2600 MW of power. These developments only commenced with the NWDT Award and are based on M.P's 18.25 MAF water allocation which it, like the other riparian states, must utilise within 45 years (2025) when the Award comes up for review.

GOMP prepared a Narmada Development Master Plan. By 1993, three major dams had been completed (Tawa, Barna and Sukta) while three others were partially complete (Bargi, Mateyari, Kolar), and five others under implementation (including Man, Jobat and Chhota Tawa). Nine major projects are being surveyed and 10 others remain to be investigated. As regards medium projects (now likely to number 158), 18 have been completed and nine are in progress. Nine more have been surveyed and 13 others are being investigated. Among small projects, as many as 1018 have been completed and 112 are in progress. The projects so far completed have created an irrigation potential of 580,000 ha and utilise 4 MAF of water. (UNI, August, 1993)

Work on M.P's mother dam on the Narmada, Indira Sagar, commenced in mid-1992 with the award of contracts. The two major dams below it in the same cascade, constituting the NSP complex, are also being taken in hand as the three go together. The NSP Complex and the Man, Jobat and Chhota Tawa projects (whose commands are expected to accommodate the R & R programmes for the NSP oustees) are to be completed in the 10th Five Year Plan (2006) along with the Bargi Diversion scheme. Hopefully, the NSP dams will reach crest level by 1998 by when 410 MW of power (ISP and Maheshwar) and 36,000 ha of irrigation (ISP) should be available. Full generation from the entire NSP cascade and 270,000 ha of irrigation should be possible by 2005, with 147,000 ha of irrigation from Omkareshwar. (GOMP, August 1992)

These same projects (inclusive of M.P's share in the SSP) were estimated to cost Rs. 5330 crores and entail displacement of about 200,000 persons in 561 villages. The NVDA budget allocation for all schemes, major, medium and minor, in the Eighth Plan is Rs. 1526 crores. Prospects for funding a Narmada Valley environmental study ($ 90 m) and ISP have receded and financial constraints will have to be overcome. Maheshwar is a purely power project (320 MW; estimated cost Rs. 465.63 crores) and the State Government is exploring the possibility of handing over its execution to the private sector.

The Narmada Valley in M.P. is narrowly confined and has a relatively small cultivable area. The NVDA has accordingly prepared plans for trans-basin diversions. One scheme envisages diversion from the Bargi reservoir by lift to the Rewa-Satna area in the Tons-Sone basin to irrigate this water short region. There is also a proposal to lift Narmada water from Hoshangabad to irrigate lands at higher elevations in Hoshangabad and Dewas districts and, subject to techno-economic viability studies, to lift water to low rainfall areas around Dhar and Mandsaur. (Varma, 1989) A more ambitious project to lift large quantities of water from Indira Sagar to the Malwa plateau in stages, utilising a later pump-storage scheme at ISP, has been considered. The costs would be considerable and the base load power surplus required for pumped-storage may in any case not be available before 2010. This means that such options will really not come on the agenda for some time. (Master Plan, September 1991)

The height of the ISP was fixed at 860 feet in order to optimise power generation consistent with delivery of regulated releases of water to SSP whose height was determined in consequence. The arrangement has been incorporated in the NWDT Award. Hence a proposal mooted by MOEF at one stage to lower the height of the dam to 814 feet, in order to mitigate displacement and environmental losses, was formally considered and rejected. For somewhat similar reasons, a lowering of the height of SSP from FRL 455 to 436 was urged before the NWDT as the additional height meant increasing the submergence area in M.P. by 35 per cent, affecting 70 per cent more people. But the NWDT ruled that a height of FRL 455 should be maintained to ensure the stipulated hydro-electric benefits. (IRM, 1992)

Meanwhile, a number of environmental and other studies pertaining to NSP have been completed and more are coming in. (NVDA, April 1989) The NSP cascade will affect 339 villages (249 of them on account of ISP) and a population of 100,000. But only 69 villages will be fully submerged by ISP. In 38 villages neither any abadi or private land will go under and in another 25 villages no abadi land will be lost and under 20 ha will be submerged. Given the possibility of drawdown cultivation and extensive fishery opportunities, many families are likely to relocate *in situ* rather than opt to move to more distant places. In any event, R & R will need to be handled with compassion and sensitivity. As in the case of SSP, power benefits from the 1920 MW installed capacity will decline as irrigation diversions increase. Thus firm power from the NSP complex will fall from around 450 MW to 130 MW. But, as in the case of SSP, actual energy production will be higher on account of larger water availability in the flush season. There will also be some flexibility for peaking operations.

The NSP has been assailed by critics for its large submergence and displacement with what are seen as modest benefits. What is missed out is that as the "mother dam", it has to be seen as part of the larger Lower Narmada Valley development programme including the NSP complex and SSP. Its trans-basin irrigation benefits are also seldom taken into account. Seen as a whole, ISP plays a key role and is certainly cost-effective. There is criticism of Narmada Valley developments on two other counts. The first relates to past performance on rehabilitation, especially with regard to the Bargi Dam. This particular charge has some force. Bargi, like some other projects, was taken up before the upgraded NWDT R & R package was evolved by GOMP. Initial planning was weak and there have been delays, confusion about enumeration of PAPs and resultant dissatisfaction which has found expression in recent agitations. While the demand for scaling up the Bargi R & R package to SSP/NSP levels may not be possible, as retroactive application of future benefits could pose embarassing administrative and political problems, the original package must be fully implemented. This is what has now been promised and joint committees of the GOMP, NVDA, the oustees and NGOs (led by Dr. B. D. Sharma of the Bargi Visthapit Sangh) have been set up at the State, divisional and sub-divisional levels to monitor implementation. Employment is one demand that could be met were canal construction to be expedited to utilise the waters stored behind the dam. Resources must be diverted to finish the project expeditiously.

The second point of criticism relates to waterlogging and problems of irrigation in black cotton soils, especially in the Tawa command. Tawa again is among the older projects (1975). The reservoir floor was not de-stumped after clear-felling, thus affecting the lake fishery. Land shaping was not too scientifically carried out in the early phases, with topsoil being turned down in many fields in the process. Lower order water channels were not lined and drains were choked with Ipoemia weeds which resulted in loss of productivity in certain villages for a while and some waterlogging. But far from waterlogging being widespread and having ruined the entire command, only 335 ha in a few villages are affected. Even this land is cropped. Efforts are being made to lower the water table through pumping and improved surface drainage.

Meanwhile, kharif cropping has spread in place of extensive fallows and soyabean, sunflower and safflower are widely cultivated. Cropping intensity in the Tawa command is around 130 per cent. Improved varieties have been introduced and yields are up. Signs of prosperity are evident. Land prices have risen. Scope for improvement remains, but to describe the Tawa project as an outright failure is very far from true. As far as black cotton soils are concerned, the NSP command is better placed as lighter vertisols are more common. Heavy and medium vertisols increase as one moves up the Narmada valley. Relatively more favourable irrigability factors and past experience should enable NSP planners to avoid the errors committed in Tawa or Barna. Conjunctive irrigation is being planned from the start.

Funding and Benefit-Cost Ratio

The Narmada Valley projects are admittedly gigantic and will entail heavy investments. While this will no doubt strain the States' financial resources to the limit, the task is not impossible and will bring its own rewards as the benefits are expected to be even greater. This assumption is, however, strongly challenged by critics who claim that costs have been under-estimated and benefits greatly exaggerated. In their judgement, the requisite finances will not be forthcoming or, if available, will skew Plan and regional priorities and result in crippling indebtedness. With the withdrawal of World Bank and Japanese funding the resource problem will be even more acute. The conclusion drawn is that the SSP and NSP are unviable and have a poor or negative benefit-cost ratio.(NBA, Medha Patkar, June 1993; Paranjpye, June 1990; Alvares and Billorey, 1988; Amte, 1989, 1990; Human Rights Campaign on Narmada, May 1993; Eminent Persons Group, June 1992; Kothari, 1984).

Planning Commission approval for the SSP came in 1988 at an estimated cost of Rs. 6406 crores at 1986 prices. On the advice of a consultancy firm, C.C. Patel & Associates Ltd., the original construction schedule spread over 17 to 22 years was compressed to 10 years for the dam, main canal, and power stations, with other works in progress thereafter, the object being to realise benefits early and curtail overheads. 1998 was set as the target date, with first irrigation and power benefits by 1995/96. With subsequent revisions, costs were placed at Rs. 9400 crores and, taking delays, devaluation and inflation into account (not constant prices) may end up around Rs. 13,500 crores — not Rs. 25,000 crores as the NBA keeps asserting despite the possibility of the SSNNL having to make additional payments for land acquisition should the courts uphold certain claims. Gujarat's net share in the project is reduced by some Rs. 1500 crores which is to be contributed by the other participating states. About Rs. 1200 crores had been spent in the Sixth and Seventh Plans and by 1993-94 total expenditure will have reached Rs. 3500 crores. There has, however, been a setback in forgoing the $ 180 m balance of World Bank assistance and a like sum promised by Japan for turbines. The Finance Minister has, however, agreed to make good the lapsed World Bank funds in foreign exchange and may be persuaded to treat the Japanese grant similarly.

In keeping with the trend of Gujarat's annual Plan expenditure and the sums allotted to irrigation and related sectors, the State expects to be able to mobilise adequate resources through innovative approaches. Rs. 200 crores have been negotiated with ONGC as an inter-corporate loan. Bonds have been floated and advance deposits sought from beneficiary farmers and corporate bodies. Political will and sacrifice will no doubt be required. But if ever popular will was manifest for a project it has been in the case of the SSP in Gujarat which the vast majority of the populace, urban and rural, see as the State's "lifeline" and consider an "article of faith".

Cost-benefit calculations are another matter of contention. These were estimated as satisfactory and, indeed, favourable by the Project authorities and Government as well as by the World Bank. Paranjpye, whose figures have been adopted by NBA and INTACH has, however, pronounced both the SSP and NSP as unviable on the basis of a series of assumptions that discount benefits, inflate costs and project time-overruns. He sees the SSP and NSP yielding internal rates of return (IRR) of no more than 9.2 and 9 per cent respectively on a 12 per cent discount rate, thus implying negative rates of return on the investments made. His B-C calculation came to 0.65.

Alvares and Billorey are even more pessimistic. They state that against a threshold B-C ratio of 1.5 laid down by the Planning Commission (though 0.7 for water resource projects in drought prone areas), the figures touted for SSP were at first 1.88 in 1982, next 1.74 in 1984 and then 1.52 in 1987. At this stage, they say, MOEF dropped a bombshell by placing the environmental losses (ecological value) at Rs. 10,000 crores for SSP and Rs. 30,000 crores for NSP. This was even communicated to the Prime Minister in a note prepared in connection with the decision on whether or not to accord conditional environmental clearance to the SSP and NSP. This aspect is separately discussed. However, Alvares and Billorey conclude that the B-C ratio of SSP is 0.38 and that of NSP 0.17 (without MOEF's environmental costs) and 0.24 and 0.017 if the full environmental costs are included. The authors, of course, make their own inclusions and exclusions in arriving at the other costs and benefits. For instance, they argue that floods might increase and that there could be a substantial loss on fisheries.

The NBA, Baba Amte and many other critics have been in full cry ever since. Hence the title of the Alvares-Billorey book *India's Greatest Planned Environmental Disaster*!

Dr. Paranjpye's assumptions were challenged. He was found to be wrong in stating that several costs had been excluded such as that of the Garudeshwar weir, the Vadgam saddle-dam, an SSP-grid transmission link, contingencies, R & R outlays and so on. And he was found to have got a decimal point wrong in his figures of compensatory afforestation costs to add Rs. 460 crores to the total. He was also faulted on his lower estimate of crop yields with irrigation and for ignoring employment benefits.

(Alagh, July 1989 - based on an earlier draft of Dr. Paranjpye's book; and Sanat Mehta, 1990).

The critics' calculus is selective, arbitrary and incomplete. Fishery and forest/plantation benefits, for example, are going to be significantly more, especially the latter. Likewise, groundwater recharge. Savings on famine relief, ferrying of drinking water, the enormous social costs of mass migration, and flood losses are simply ignored or possibly not even understood. Benefits such as the incremental supply of fodder and fuelwood in the command, employment, peaking power and public health gains are unnoticed. Both Paranjpye and Alvares's calculations are caricatures of reality.

A World Bank handout answered many of Dr. Paranjpye's doubts. (December 1990) It noted the SSP's central role in alleviating poverty and an increasingly oppressive drought-cycle in Gujarat. It found that the benefits from catchment area treatment, which are significant, had been left out. It saw no way Gujarat could cope without augmenting irrigated agriculture and saw no alternative to the SSP, whatever supplementation small schemes might offer. It established the inclusion of compensatory afforestation costs as well as the cost of submerged forests to be a case of double counting. It could not agree with the arbitrary reduction of 20 per cent irrigation benefits on account of unsuitable soils after having included the cost of taking irrigation to difficult areas. It detected considerable confusion between reference to economic and financial costs and between nominal and real rates of interest. The dismissive attitude to power benefits was very curious as also allowance for no more than a 40-year drinking water supply on the (totally unfounded) argument that water availability would have run out by then, if available at all in the first place. Finally, the Bank envisaged a "without (SSP) project" scenario as entailing either massive recurrent relief expenditures or creating potential "oustees of the future" who would take to the streets.

MOEF's Environmental Legerdemain

The role of the Ministry of Environment and Forests in delaying and obstructing Narmada Valley development in the earlier years cannot go unnoticed. It exhibited all the zeal of the new convert in withholding environmental clearance for the SSP and NSP for a full four years (1984–87). It went back on a solemn GOI commitment written into the SSP loan agreements with the World Bank in 1985 to release degraded forest land for R & R. This resulted in huge delays, uncertainty, cost escalation and loss of project and official credibility with tribal PAPs until 2700 ha of Taloda forest land was finally released in 1990. To plead violation of the Forest Conservation Act of 1980 is specious as a firm prior commitment to release degraded forest land had been made to the NWDT and incorporated in its Award, a judicial order. Moreover, the Forest Conservation Act was well in place when GOI signed the loan agreements in 1985. The paranoid over-centralisation of regulatory powers by MOEF under the Forest Conservation Act without due discretion went to such extremes that all Chief Ministers had ultimately to join forces to put an end to this demonstrably counter-productive regime. Meanwhile, elements even of the celebrated Chipko movement turned anti-forest, such was the rage at mindless impositions.

Conditional environmental clearances were granted to SSP and NSP in 1987 on the basis of many unreal and sterile deadlines which provided a stick to all and sundry with which to beat these projects. The MOEF briefing note formally sent to the Prime Minister in 1987 showed strong bias against the Narmada projects on alleged grounds of inadequate data, their poor state of readiness, unacceptable environmental costs and lack of water availability which was stated to be 23 MAF as against the design estimate of 28 MAF. The land submerged by the NSP was placed at over 91,000 ha of which a little over 40,000 ha was forest land (the corresponding forest submergence under SSP being 13,700 ha). MOEF argued that since only 141,000 ha would be irrigated by the Indira Sagar dam, the same amount of irrigation could be provided by reducing the dam height from FRL 860 to FRL 814 which would reduce forest submergence dramatically to 8075 ha and the affected population of 129,396 (1981 census) to 20,020. This completely ignored the drastic impacts on storage regulation, irrigation benefits from Omkareshwar (135,000 ha) and proposed lift schemes, and power benefits in the entire NSP complex, and suggested nothing in their place. As much as 7.75 MAF of storage was calmly written off.

NSP planning was seen to be trailing behind SSP and hence MOEF argued that "it is neither desirable nor recommended that the SSP should be given approval in isolation on technical and other grounds" until the NSP "Intention Plan" is converted into an "Action Plan"

The prize statement in the MOEF note was its estimating the forest loss of NSP at Rs. 320 crores (with no corresponding figure for SSP) and the environmental cost of loss of forests at Rs. 30,923 crores and Rs. 8,190 crores respectively, thus rendering both projects totally unviable at a stroke. The environmentalist lobby in India and abroad, gleefully clutched this new theorem to launch a tirade against the Narmada projects, labelling them the world's greatest planned environmental and financial disaster.

Writing on the identical theme four years later, a senior Adviser in MOEF, reiterated the very same thesis. (Maudgal, January 1991) It was his contention that the NSP and SSP had generated "unprecedented hostility not only among the affected population but also in environmental groups and NGOs throughout the country. Nature lovers from other parts of the world have also registered their shock and concern". He described the environmental impacts of both projects as "massive and serious" and wondered if the ameliorative action proposed would be successful. If it failed "it will prove to be an unmitigated environmental disaster". No word about project benefits!

Examine the "environmental cost" thesis. On what does it rest? What is the evidence or basis of calculation? The computation supposedly rests on the sum total of all ecological benefits — water inflitration, humus, oxygen bank, carbon sink, holding the soil, providing niches and habitats for a variety of floral and faunal species and micro-organisms — expected from a well-stocked forest with a density of 10 over a 50 year period at the rate of Rs. 126.7 lakh per ha. Much of the Narmada Valley forests coming under submergence are degraded and the average density is estimated at 0.4. So even if the MOEF's formula were to apply, the so-called environmental cost of forest loss would only be 25 per cent of what has been indicated.

The basic fact, however, is that the MOEF calculation is unsubstantiated and unscientific. This is revealed in an interview with Dr. S.P. Pant, Evaluation Officer, NVDA. He asked to know the basis of MOEF's formula and was simply told that this was a given figure. When he argued that a corresponding environmental value should then be placed on the new afforestation and plantations being undertaken after, say, a 10 year maturity period, "the Department of Environment would not accept that" and would give no reason: "I went so many times. . . I tried to give what I felt is the correct judgement. They said, 'No', we don't need it". (Alvares and Billorey, 1988) Very revealing.

This precisely is the catch. For every tree being lost to the Narmada projects, vastly more numbers are being and have already been planted. The so-called environmental benefits from these new forests and plantations, even if discounted, will undoubtedly be many times more than the alleged environmental loss thus yielding an immense positive benefit ratio in favour of the SSP and NSP.

Indeed, from the global warming point of view, the carbon/oxygen balance as a result of the Narmada projects, and certainly the SSP, will improve vastly on account of the irrigated crops and tree cover in the command area. No benefit on this account has been claimed by the project authorities and, characteristically, none has been admitted, let alone allowed, by MOEF. The World Bank at least has included a notional net gain on this account merely to indicate its presence.

Exit the World Bank and Other Donors

The IRM Report was flawed and perverse. (ARCH, April, July 1993) It was biased, selective and quite out of tune with technical, economic, political and cultural realities. It, maybe subconsciously, adopted North American standards. It is interesting that Berger, the co-Commissioner, and the two senior advisers, Donald Gamble and Hugh Brody, all Canadians, worked together on the Mackenzie Valley Pipeline Inquiry relating to an Arctic gas pipeline project and were all closely involved with Canada's Department of Indian Affairs and Northern Development. Berger was instrumental in the inclusion of aboriginal rights in the new Canadian Constitution. In attempting to transplant this experience of wide open spaces and a miniscule indigenous population to India in the context of SSP, the Mission developed an instant sociology, an instant anthropology, an instant history, a novel cultural framework and a new federalism and managed to get the Indian reality all wrong. It did not address its terms of reference. It self-confessedly went far beyond it. Not that Indian planning or the SSP and NSP are without blemish in terms of policy, design or implementation. The IRM did offer some useful insights and could have made recommendations to improve the projects and their implementation. That would have been a real contribution. Instead, from small premises it leapt to magisterial judgements: rehabilitation is impossible; the project environmentally unworkable. No hint of doubt or any appreciation of project benefits or the cost of further delaying or abandoning the project for that is what its advocacy of "stepping back" might well have implied as the critics prepared for the kill. The aspects of Indian environment that the IRM so determinedly tried to safeguard are precisely those that are being further degraded and destroyed by lack of development and demographic pressures.

India perhaps made a mistake in accepting an Independent Review Mission in which it was unrepresented, in the choice of its members, over which it sought no say, and in permitting it to set forth on a fishing expedition. It expected better. The World Bank reacted with surprise as its constructive efforts and genuine concern with ensuring a positive outcome through a sensible incremental, *pari passu* approach was turned against it and it was put in the dock. If the World Bank has in 45 years learnt and innovated development economics and lending, much of that has come from its rich Indian experience. And the Narmada Valley projects could have been another international landmark.

However, the Bank pointedly rejected some of the basic conclusions of the IRM both on R & R and environmental issues and showed disinclination to "step back". It admitted possible lapses but justified its broad stance and suggested certain "next steps" to re-examine premises and progress and set right whatever was found wanting. It was for continued support to the project against a set of performance benchmarks to be followed by an evaluation within six months.

Bradford Morse and Thomas Berger were outraged. Their mission over, they nonetheless set out to revive the Mission by writing to the Bank President and actively lobbying Executive Directors to impose their agenda on the Bank. They could not wait for Directors to read their full report and form their own judgement. They summarised what they thought the Directors should know and discuss. Nobody protested. Busy Directors, not too well informed of the details and representing countries with strong or emerging Green movements, and beginning to take positions for the forthcoming UN Conference on Environment and Development, thought fit to strike a blow for Rio.

No surprise then that before the "probationary period" was over, GOI informed the Bank that it was withdrawing SSP from the Bank's lending portfolio, giving up an unspent $ 180 m. It instead decided in consultation with the four concerned State Governments to complete the projects without external assistance. The Gujarat Assembly adopted an unanimous resolution on July 16, 1993 accusing the IRM of going beyond its terms of reference and behaving as "an advocate for the activists". It rejected the Mission's findings and said it would go ahead with the Project. The European Parliament adopted a somewhat uninformed resolution on July 9 not merely endorsing the IRM Report but calling on the Bank to withdraw from the Project and GOI and the State governments concerned to stop further work on the SSP and "seek and implement viable alternative means of supplying water to areas threatened by drought". The Gujarat Assembly sent back the text of a resolution it adopted expressing "unhappiness and resentment" at the European Parliament adopting a position without caring to hear the other side.

What do the Critics Say?

The critics see virtually nothing good in the SSP and NSP. Both projects are seen as mistaken and unworkable. As the Sardar Sarovar dam kept steadily rising and the first submergence approached, the NBA announced the struggle had reached a critical point and prepared for a last ditch battle. Hence the Manibeli satyagraha, the threat of *jal samarpan* and Medha Patkar's indefinite fast. Over time, however, the cause and context have undergone change.

At first, the struggle was for better terms of R & R. As soon as real progress began to be made in that direction, this issue was dropped and attention turned to the alleged violation of the Forest Conservation Act at Taloda; then to lack of data and "disastrous" environmental costs; later to questioning all benefits; then to the inviolability of tribal rights and culture; on to issues of funding and financial viability; still later to equity for Kachchh and Saurashtra; and, finally, always there but now loud and clear, the denunciation of the present national development strategy and the demand for a new socio-political paradigm. Contradictions, misconceptions and muddled facts have not diluted faith. Nor the steady falsification one by one of several basic assumptions about the allegedly crippling effect of inadequate data or lack of studies and, particularly, of the impossibility of getting a fair R & R package, of finding land, of degraded forest lands being released, of persuading people to move. As Bank officials also noted, the NBA's criticisms and even objectives kept changing. When R & R seemed in "danger" of working, the slogan changed to "stop the dam".

One has not only to live for the day but safeguard the future. A Bank handout explained how the numbers threatened by drought in Gujarat has increased sharply, and relief costs with that; how groundwater levels and quality have deteriorated; how population in the command area will increase by 8 m by 2020. Looking ahead at a scenario of "increasing public expenditure (on relief) and escalating hardship", these people could be seen as "resettlers of the future. . . They do not hold public meetings now but in the year 2020, if you deprived them of their Sardar Sarovar dam, they would certainly be out in the streets". (World Bank, April 1990) The Bank felt that considerable progress had been made in achieving resettlement policies that were arguably "the most favourable in any developing country". An independent consultant similarly opined that "data collection and collation are well advanced and very few projects in developing countries have so much data gathered on such a wide variety of subjects". (Wallingford, March 1993)

The NBA's latest objectives and those of Baba Amte focus more on development and technical alternatives. (NBA, 1991-92) Baba Amte, like Paranjpye and Alvares, has proposed alternatives to the current "large-scale cultural ethnocide". Myth and emotion rule strong. Medha Patkar laments loss of "rain forests" in M.P. (Where are they?) She sees "a powerful coalition of rich farmers, industrialists and contractors" rooting for the dam in order to fill their coffers. The NBA, in turn, claims that the basic assumptions of the NWDT have gone wrong such as the Maximum Probable Flood. (Who said so? This is just not true.) The sheer verbal extravagance of some of the critics can be amusing though utterly misleading. Thus: the SSP will submerge "one of Asia's largest unsurveyed forests. Quantities of medicinal plants and flowers. . . will be drowned"; "the single greatest user of precious water will be the water-profligate chemical industry such as the ones in Baroda"; the tribals in Manibeli "even manage without the ubiquitous safety match, so wasteful of timber." They make fire "by striking flint against the floss of the silk-cotton tree. What are we offering them? Pollution, prostitution, Aids"; "the Narmada forests are still largely unsurveyed for minerals like gold, silver, gas, oil and uranium which have never seriously been tapped". (Bamroo, May 16, 1992) Much of this hallucinatory prose leaves one between tears and laughter.

Here are Their Alternatives

The NBA wants another development path. Others have suggested alternatives. Let us look at them. Alvares and Billorey have three: lower dam heights, (but what of the loss of benefits?), river-lift irrigation as in West Nimar, (but if this is economical why should farmers along the foreshore not continue to practise it?); and small, single-purpose, localised reservoirs in cascades, (but how is this being prevented where feasible?). These are not alternatives as much as complementary strategies if and where feasible.

Vijay Paranjpye has another set of alternatives. These include small water harvesting programmes in the tribal districts; selective low-head medium dams; and, again, river or canal lift irrigation through integrated watershed development. There is, Paranjpye informs us, every prospect of Kachchh being able to treble food and fodder production by tapping local sources of surface and ground water. This may seem good news to all except those who have been desperately trying this for years. Some progress is indeed being made in this direction. But the harsh fact is that all small projects in Kachchh and even in less arid regions fail in drought years and in critical periods when most urgently required. The notion that Narmada water cannot and will not reach Kachchh and, even if it does will dry up in 40 years, is without foundation. For SSP power, Paranjpye suggests a natural gas option. But why as a replacement and not as an additional (though depleting) source to the extent available? And if Gujarat has 63 billion tonnes of "underground coal reserves which could yield 15,000 billion cu m of natural gas which would be sufficient to meet not only the energy requirements of Gujarat but also large requirements in the States of Maharashtra, M.P. and Rajasthan" this is very heartening. The only catch lies in "expeditious R & D in the field of gasification". If and when that day comes, this very deep, non-gassy coal could of course be tapped. The alternatives for Saurashtra that Paranjpye proposes are equally futuristic and iffy. (June 1990)

Baba Amte's alternatives span conservation, greater water-use efficiency, completion of on-going projects, all of which suggest better housekeeping on which there can be little disagreement. He too then goes on to propose small dams, watershed management, improved dryland farming techniques, a total catchment management approach for Kachchh to satisfy all its water requirements, bio-mass energy and energy conservation.

The Group of Eminent Persons who wrote to the Prime Minister also advocated "integrated micro-watershed development" with small check dams and afforestation for Kachchh which have been successful in drought-prone areas of Maharashtra and Karnataka. But similar micro-projects are being developed in Kachchh and Saurashtra. And everybody seems to forget that whole villages by the score are abandoned and people migrate in lean years in search of water because there is little or no rain. The SSP Kachchh Branch and compensatory afforestation scheme will bring the water and promote the forests/ plantations for which there is all this yearning.

The NBA scoffs at the fanciful idea of SSP providing irrigation or drinking water to Kachchh, warns of waterlogging and salinity should the water actually get there, and then criticises the project design for dedicating too little water for Kachchh and Saurashtra. This is having in both ways. To its way of thinking, "groundwater is the real lifeline of Gujarat" (June 29, 1993). The NBA appears to be unaware of the groundwater situation in North Gujarat, Saurashtra and Kachchh and places an unrealistic premium on large scale recharge through watershed management that will suffice to insure even against a single drought.

An even more visionary alternative is to deepen river channels in Saurashtra and Kachchh in some stretches to "create vast captive river lakes from which solar-powered pumps can lift water and send it by pipeline to nearby and middle distances". (Vasu, May 16, 1992) Here is yet another intellectual mirage.

All these "alternatives" are fantasies or are supplementary devices which could certainly have a place where and if techno-economically feasible. It is nobody's argument that the SSP and NSP will exhaust all need for further development. Much more will still remain to be done and complementary programmes must be pursued. The other notion that small, medium and large dams and groundwater are mutually exclusive is equally mistaken. Each has its place and they can and often do work in tandem. The very critics who tirelessly flay the giganticism of the Narmada Valley development programme, start by mentioning that the overall project concept envisages 30 large dams, 135 medium dams and 3000 small dams. There would probably be many thousand more micro projects along the line such as the deepening and replenishment of tanks and ponds alongside the canal network, innumerable catchment area treatment check dams, gully-plugs, detention ponds and the like. And, unlike any other irrigation project so far, conjunctive use of groundwater and the sweetening of saline aquifers is part of the project design. In other words, the SSP and NSP aim to do all the critics pose as "alternatives" on a far more massive scale than they suggest and this is simply not understood or is glossed over.

The small dam versus big dam debate is thoroughly misconceived and rests on unscientific premises when sought to be stretched and made an absolute principle. Small dams are not cheaper nor do they submerge less land than large dams. On the contrary, they submerge far more land and by and large cost far more per unit of irrigation over a given life. Submergence is a simple matter of geometry. And the small life of small dams and their proneness to run or remain dry when their limited catchment suffers drought needs no remarkable proof. Their shorter life and, otherwise, high maintenance cost per unit of water is also well established. The low cost of tubewell and lift irrigation supplies would be greatly enhanced were large energy subsidies not concealed.

A comparative exercise was done by the Central Water Commission to substitute the Heran dam, a large dam, in the lower reaches of a river in Saurashtra by six smaller dams in the upper reaches of the river. The results were revealing, with the single dam scoring over the six-dam option (figures in brackets) in all parameters. Thus, gross storage 380 cu m (230); irrigation capacity 39,500 ha (28,700); gross submergence 3503 ha (3650); affected villages 20 (21); affected population 8270 (10,300); and capital cost Rs. 19.70 crores (25.48 crores). A second investigation made by CWC of the proposed Girna dam in the Mahanadi basin in Orissa versus an option to build a smaller Girna dam plus 8 satellite storages to provide a comparable storage capacity of 603 m cu m, indicated a 150 per cent higher cost, 60 per cent more submergence, greater evaporation losses, and lower power generation in the case of the nine-dam option. (Mistry, December 2, 1991)

In sum, the alternatives offered submerge more land, are more expensive, less cost-effective, less reliable, or plainly futuristic or visionary. They are at best complementary programmes which are in any case being pursued and are not mutually exclusive.

As far as lowering the SSP or NSP dam heights or planning another configuration of dams in the Narmada Valley are concerned, it must be understood that each of the State Governments had considered several options and permutations and combinations over a quarter century. The NWDT weighed all these, considered various combinations of dams, heights and other parameters and arrived at the present configuration as defining an optimal solution. True, the Award can be revised by mutual agreement even within an otherwise binding 45-year period. But why should an optimal solution be undone for a sub-optimal package in the absence of any other compelling factor that would not confront the new package? Merely lowering dam heights at the cost of major benefits is to beg the question.

Every Reason to Hasten

The Narmada Valley has waited long to see its waters harnessed. Now that that day has come with a resolution of inter-State disputes and the SSP and NSP can deliver not merely regional but truly national benefits, let them not be mindlessly delayed or stalled. The twin projects, and the more advanced SSP in particular, have been carefully designed to exacting technological and management standards with R & R packages that are part of path-breaking national and globally-evolving norms. The projects per se involve no violation of human rights, neither national law nor the ILO Conventions 107 nor the latest protocol on forced evictions. There was an initial sluggishness in evolving the present R & R package and in keeping people informed so as to eliminate uncertainty. That phase is over. Environmental standards are being met in good time despite unreal demands and unrealistic deadlines earlier imposed, and no significant or irretrievable injury has resulted.

The R & R norm is that at the end of the day, every displaced person should be at least as well off if not better off than before. This is already beginning to happen and it is singularly unfortunate and grossly unfair that transient problems have been highlighted as permanent outcomes. Rehabilitation is in any case a process and not an instant event. It means "to invest with dignity", a truly uplifting and humane concept which must be the guiding spirit of R & R in the years to come.

The critics, including the IRM, have been systematically proved wrong or exaggeratedly concerned as steady improvements have been effected and earlier problems gradually overcome. Apocalyptic deadlines have passed without any catastrophic consequences while the objectives and practical schedules have by and large been fulfilled in great measure. The incremental and *pari passu* strategy has paid off, contrary to the laboured reasoning of the IRM which, unfortunately, could not rise above its own bias and preconceptions.

Many problems remain and there is a long and difficult haul ahead as the SSP and NSP are huge, complex projects. NGOs have a critical and important role to play as ARCH-Vahini, Anand Niketan and others have shown. Without their vigilance, perseverance and community leadership, things may not have gone as well as they have. The project and the nation owe them much thanks.

Were it that the same could be said of other NGOs and critics like the NBA and Baba Amte. They have been carping rather than constructive critics. The NBA and its leader, Medha Patkar, have sadly gone astray. The NBA has changed tack time and again and has stultified itself. Its movement has not been without elements of coercion and passive violence. If the current "review" discussions it is having on the SSP with an Eminent Persons Group persuades it to adopt a more constructive path, that effort will have proved worthwhile. These discussions, the Government has stated, do not constitute a review of the project but could lead to recommendations for improvements or corrections for which the door has always been open. A review per se, on no new grounds, would open a Pandora's box and reduce governmental, planning and parliamentary processes to naught.

The Narmada Valley projects represent a great challenge and opportunity. They can be transforming. They are part of a better tomorrow that cannot be cancelled and must not be further postponed.

REFERENCES

1. Narmada Water Disputes Tribunal, Report of, Government of India. Vols. I, II, New Delhi, 1978 and Further Report, New Delhi, 1979.
2. Narmada Planning Group, Planning For Prosperity: Sardar Sarovar Development Plan. Sardar Sarovar Nigam Ltd., Gandhinagar. November, 1989.
3. Paranjpye, Vijay, High Dams On The Narmada: A Holistic Analysis of the River Valley Projects. Studies in Ecology and Sustainable Development. INTACH, New Delhi, June, 1990.
4. Independant Review Mission (IRM) on Sardar Sarovar by Bradford Morse and Thomas Berger. *Resources Futures International*, Ottawa, June 1992.
5. Verghese, B.G., *Waters of Hope*. Oxford-IBH Publications Ltd., New Delhi, 1990.
6. ARCH-Vahini, Are The Critics Right- R & R in SSP. Mangrol, Via Rajpipla, Gujarat- 393 150. April 30, 1991.
7. Multiple Action Research Group (MARG), Sardar Sarovar Oustees in Madhya Pradesh: What Do They Know? Vasudha Dhagamwar, Director, and Enakshi Ganguly Thukral, Project Coordinator. New Delhi, May 1987.
8. Narmada Control Authority. Salient Features of R & R Policies for SSP Affected People. Indore. June 1993.

9. Sardar Sarovar Punarvasavat Agency. Rehabilitation of PAPs of the SSP, Indore, June 1993.

10. Narmada Valley Development Authority, *Rehabilitation Policy for the Oustees of Narmada Projects*. Bhopal, 1992.

11. ARCH-Vahini. *Morse-Berger Review: A Search For Truth*. P.O. Mangrol, Via Rajpipla, Gujarat - 393 150. November 8, 1992.

12. Narmada Shikshan Samiti. Memorandum from PAPs of SSP of Akrani Taluka in Maharashtra and Arch-Vahini, Gujarat to Chief Minister of Maharashtra, Mantralaya, Bombay. (Translated from the Marathi text). April 29, 1992.

13. *The Lawyers*. The Battle in the Courts (Cover story). Bombay, July 1991.

14. Alvares, Claude, Surveyor of Doom. *The Illustrated Weekly of India*, Bombay, January 5-6, 1991.

15. Tata Institute of Social Sciences. Resettlement in the Taloda Forest Area: a Study of the Area and its Inhabitants. Monitoring and Evaluation of R & R of SSP Displaced in Maharashtra. Report No. 14. Bombay, May 1991.

16. S. Parasuraman, Nandini Rao TISS Group. Sardar Sarovar Project, Review of R & R in Maharashtra by Ajith Kumar, A. Ahire, J. Gawale, D. Mali, R. Mohite, *The Economic and Political Weekly*, Bombay, August 21, 1993.

17. ARCH-Vahini Morse-Berger Review of SSP : A counterview. Mangrol, Rajpipla - 393 150. January 9, 1993.

18. Varma, S.C., *Human Resettlement in Lower Narmada Basin* NVDA, Bhopal, July 1985.

19. Alvares, Claude and Billorey, Ramesh. *Damming The Narmada, India's Greatest Planned Environmental Disaster*. Natraj Publishers, Dehra Dun, 1988.

20. Narmada Valley Development Authority, GOMP. SSP: Action Plan of R & R for the Oustees of M.P.Bhopal, March 1993.

22. Amte, Baba. *Cry, The Beloved Narmada*. Maharogi Sewa Samiti, Anandwan, Chadrapur District. July 1989.

23. Dhule District Court. Registered Civil Suit No. 33/1992. Order of Civil Judge, Mr S.D. Gundecha. Dhule, March 3, 1992.

24. Fact-Finding Team appointed by Ministry of Water Resources, GOI., Report of Incidents that occurred in Manibeli village and its vicinity between March 26 and April 4, 1992. Government of India, Delhi, April 1992.

25. *Times of India* News Service. "Private probe reveals police atrocities in Manibeli". Baroda, April 24. *The Times of India*, Bombay, April 25, 1992.

26. Dhagamwar, Vasudha. "Samarpan or Suicide? It is the Move-

ment, not Personalities that Matter". *Indian Express*, Delhi, September 4, 1993.

27. Supreme Court, Writ Petition (Civil No. 1201 of 1990. Dr B.D.Sharma, Petitioner, *vs.* Union of India and Others, Respondents.

28. Supreme Court, Writ Petition (Civil) No. 1201 of 1990. New Delhi, August 9, 1991.

29. Asia Watch. *Before the Deluge: Human Rights Abuses at India's Narmada Dam.* Volume 4, Issue 15. Washington, June 17, 1992.

30. Mohanty, Manoranjan, Human Rights Campaign on Narmada:Campaign Statement on the Human Rights Dimension of the Narmada Struggle. New Delhi, May 28, 1993.

31. World Bank, India-Sardar Sarovar (Narmada) Projects. Agriculture Division, New Delhi. Handout, April 1990.

32. Baviskar, Amita, The Researcher as Pilgrim. *Lokayan Bulletin*, New Delhi, May-August, 1991.

33. Dubey, K.C., The Bhil Track: A Study of Displaced Tribals in 17 Villages, SSP. NVDA, Bhopal, March 1993.

34. Ninan, Sevanti. Narmada Dam: One high-profile crusader takes on another. *Down To Earth*, New Delhi, April 30, 1993.

35. Indian Institute of Management, Centre for Management in Agriculture. Compensation to Canal-Affected People of Sardar Sarovar Narmada Project. (Sponsored by SSNNL). U.K.Srivastava, S.P.Seetharaman and S.S.Mehta. Ahmedabad, October 1991.

36. Sardar Sarovar Nahar Nigam Ltd. SSP Water Delivery and Drainage - Canal Oustees. Gandhinagar, 1992.

37. Joshi, Vidyut, *Rehabilitation —A Promise to Keep:(A Case of SSP).* Tax Publications, Ahmedabad, June 1991.

38. High Level Group, Government of Gujarat, Sardar Sarovar: Initial Response to Independent Review Report. Gandhinagar, July 1992.

39. NVDA, Report of the Committee on Guidelines for Revision of the Master Plan for Development of the Water Resources of the Narmada, Bhopal, September 1991.

40. Government of Gujarat. Comment on the Report of the IRM on SSP. Gandhinagar, August 1992.

41. Dewan, M.L., Report of Inter-Departmental Committee on Soil Conservation and Afforestation in the SSP and NSP Narmada Catchment. Soil and Water Conservation Division, Dept of Agriculture and Cooperation, Govt. of India. New Delhi, August, 1985.

42. Revised Action Plan on Compensatory Afforestation for NS and SS Multipurpose River Valley Pojects. NVDA, Bhopal, December 1986.

43. Conservator of Forests, Office of. Statement on CAT and Compensatory Afforestation in M.P. NVDA, Khandwa, February 1992.

44. Narmada Control Authority. Status Report on Environmental

Aspects of Sardar Sarovar and Narmada Sagar Projects. Indore, July 1992.

45. World Bank, Overview of Environmental Assessment, Sardar Sarovar and Narmada Sagar Projects, Narmada River Basin, India. Don. W. Levenhagen, Consultant. March 7-8, 1991.

46. Raj, P.A., SSP: Some Technical Issues. Paper presented at Seminar on Major Issues Relating to the SSP, Sardar Patel University, Vallabh Vidyanagar, March 7-8, 1991.

47. Central Water Commission, Hydrology Studies Organisation: Sardar Sarovar Dam - Backwater Profile Assessment. New Delhi, 1992(?).

48. Sabnis, Prof. S.B., Eco-Environmental and Wildlife Management Studies of the Narmada Basin in Gujarat. Paper presented at Seminar on Major Issues in relation to SSP at Vallabh Vidyanagar, March 7-8, 1991.

49. M.S.University, Eco-Environmental Studies of Sardar Sarovar Environs. Environmental Series-2. Baroda, October 1992.

50. M.S.University, The Sardar Sarovar Narmada Project Studies on Ecology and Environment, Baroda, July 1983.

51. Central Inland Capture Fisheries Research Institute, ICAR; Sociological Survey of Fishing Families of the Narmada River. Barrackpore, June 1991.

52. Wallingford, H.R, Environment Changes Downstream of Sardar Sarovar Dam. Report EX 2750. February, 1993.

53. Dave, K.M. Chief Engineer, CAD, SSNNL. Some Aspects of Water Management in the Command Area of SSP. Gandhinagar, 1991(?)

54. Narmada Planning Group, SSP Command Area Environmental Impact Assesement. Interim Report Vols. I, II and III. Gandhinagar, April 1993.

55. Dhawan, B.D., Conjunctive Irrigation:Sardar Sarovar. *Financial Express*, September 10, 1990.

56. Wallingford, H.R, SSP Command Area Environmental Impact Assessment. Progress Report EX 2766 Volume I, March 1993.

57. Raj, P.A., *SSP: A Dip in the Drip. Thinking Aloud*. SSNNL, Gandhinagar, 1989 (?)

58. Narmada Control Authority; Benefits to Saurashtra and Kachchh Areas in Gujarat. Indore, March 1992.

59. Parikh, Dilip H.; Vora, Ambrish J. Alagh Yoginder K. Redefining Gujarat's Irrigation Potential: A Model. *Economic and Political Weekly*, Bombay, June 26, 1993.

60. Narmada Control Authority. Need for Hydro-Electric Power Development at SSP. Indore, July 1990.

61. Consulting Engineering Services (CES), New Delhi and Haskoning, Nimegan (Netherlands). Navigation on the Narmada River from Hoshangabad to the Sea. Feasibility Study Phase I: Methods to pass

Four Major Dams. Presentation report. Ministry of Shipping and Transport, New Delhi, April 1985.

62. United News of India, 21 Projects completed under Narmada Scheme. Report from Bhopal in *The Pioneer*, New Delhi, August 2, 1993.

63. Sahai, Shitla. Minister of Irrigation, GOMP; Narmada Valley Development: A Boon to M.P, Gujarat. Text of a newspaper interview, Bhopal, August 19, 1992.

64. Varma, S.C. Unharnessed Flows the Narmada into the Arabian Sea. Paper given to author. Bhopal, July 20, 1989.

65. Master Plan for Development of Water Resources of the Narmada, Report of Committee on Guidelines for Revision of. NVDA, Bhopal, September 1991.

66. Narmada Valley Development Authority, Indira Sagar Project, Brochure on. Bhopal, April 1989.

67. NBA, Why is it Necessary to Review SSP? Statement circulated at discussion meeting called by Minister of Water Resources, New Delhi, June 29, 1993.

68. NBA, A Plea to Review Sardar Sarovar. *The Other Side*, New Delhi, May 1993.

69. Patkar, Medha. All Legitimacy Stands Destroyed. Interview in *Frontline*, Madras, May 7, 1993.

70. Amte, Baba. The Case against the Narmada Project and the Alternative Perspective. Warora, Maharashtra, March 1990.

71. Human Rights Campaign on Narmada, Statement on the Human Rights Dimension of the Narmada Struggle. New Delhi, May 28, 1993.

72. Eminent Persons Group, Memorandum to Prime Minister signed by Nikhil Chakravartty, Dr Satish Dhawan, P.N.Haksar, Justice Krishna Iyer, Dr Amulya, K.N. Reddy. New Delhi, June 24, 1992.

73. Patkar, Medha. Interview in *The Lawyers*, Bombay, July 1991.

74. Kothari, Ashish and Bhartari, Rajiv. Narmada Valley Project, Development or Destruction? A Report by Kalpavriksha Environmental Action Group. New Delhi, 1984 (?)

75. Alagh, Y.K. Enclosure entitled "Some Facts" with letter addressed to Baba Amte. Planning Commission, July 5, 1989.

76. Mehta, Sanat, Chairman SSNNL, SSP - Boon to Gujarat, in SSP, A Correct Perspective, NPG, SSNNL, Gandhinagar, March 1990.

77. World Bank. Comments on the book, *High Dams on the Narmada* by Vijay Paranjpye. New Delhi, December 17, 1990.

78. Maudgal, S., Adviser MOEF. Environmental Impacts of Water Resources Development Projects in India. Paper presented at Regional Workshop on Environmental Impact Assessment for Water Resource Projects sponsored by MOEF and Central Board of Irrigation and Power, New Delhi, January 1991.

79. ARCH-Vahini, The SSP Withdrawn from the World Bank. Mangrol, March 1993.

80. SSP, An Intellectual Fashion. ARCH-Vahini. Mangrol, July 1993.

81. World Bank, India- Sardar Sarovar (Narmada) Projects. Handout. New Delhi, April 1990.

82. Bamroo, Sanjiv, "Watershed". Cover story in *Illustrated Weekly of India*, Bombay. May 16-22, 1992.

83. Vasu, M. Venishankar, There is a Way. Excerpts from a paper on water utilisation and the Narmada Valley Project. *Illustrated Weekly of India*, Bombay. May 16-22, 1992.

84. Mistry, J.F., Theme paper for Symposium on Large Dams versus Small Dams organised by Central Board of Irrigation and Power and SSNNL. New Delhi, December 2. 1991.

6

Winning the Future

IT is no surprise that all the major parties contesting the Delhi elections in the autumn of 1993 should have included water in their manifestos. A news report recalled increasing stress over this issue with "pitched and pitcher battles" over water and resultant deaths during the year. The grim scenario, it warned, is worsening. Tubewell supplies have reached their limit, with an underlay of brackish water. The Bhakra system meets some of the Capital's demand but the current 20 per cent deficit in supplies will only be made good and the future safeguarded with prospective diversions from the Tehri, Kishau and Renuka dams on the Ganga and Yamuna in Uttar Pradesh and Himachal over the next decade. (*The Pioneer*, October 1993) Distress calls were coming in from Saurashtra too with reports of 294 villages being supplied water by tankers since last year, hundreds of check dams, ponds and tanks dry, and cattle-breeders beginning to migrate once again to greener pastures in central and southern Gujarat. (*Times of India*, October, 1993)

Mounting Water Stress

Not a week passes during the long dry months without reports of drought, distress, drinking water scarcity, migration, disease and loss of production and employment in one part of the country or other on account of water problems. The monsoons do come, erratically even within their appointed season and, whoosh, the water is gone. The spatial distribution is highly skewed and the limited precipitation is insufficient to meet the requirements of a burgeoning population, livestock included, let alone permit an improvement in the quality of life which in turn would hasten a demographic transition and moderate future numbers. The sheer pressure of population is degrading the environment, aided by unsustainable survival strategies and, admittedly, avarice, to keep the vicious circle turning.

Unfortunately, too many people seem not to know it, but India is increasingly becoming a water-short country, with the situation already acute in several cities and regions and in certain seasons. Things are not going to get any better and social and economic stability and ecological security may be seriously undermined unless purposeful action is taken before the crisis deepens and gets altogether out of hand. The Worldwatch Institute has focussed on global water scarcity (Postel, 1992) and has advocated water-use efficiency, conservation, recycling, re-use, and improved water and watershed management (including resort to sprinkler and drip irrigation) for sustainable agricultural, industrial and urban use. The point is well taken, except that it will need investments, human development and technological change over a period of time before such systems are able to deliver. When they do, their effects will be important but selective and limited in most developing societies which are the ones faced by the population explosion, semi-arid or tropical climates that accentuate water stress, and have to create the necessary infrastructures. This is not to dismiss the conservation-efficiency route but to emphasise that while this is necessary it is by no means going to be sufficient in the near and medium future at least. Trying to bridge the gap by selectively turning away from obvious and available solutions, which include large dams and transfers, on doctrinaire grounds would be to invite trouble without saving the environment.

It is for the same reason that harking back to the past, which has useful lessons and much traditional wisdom to offer, may give little relief, if any. It was different in my childhood or in my father's time, is a phrase heard often enough. A return to that age is, however, foreclosed as population has doubled and tripled since and large cities and conurbations have developed which can no longer be serviced by traditional methods. Water quality has become increasingly important and the quality of energy also matters though this is not generally realised when large-scale reliance on simple, age-old or non-conventional sources are instantly advocated. All these have their uses and some technologies will surely become viable and attractive in the future. But in the long run we could all be dead. Malthus was alive and well in our childhood or father's time when malaria and famine took care of excess numbers. There can be no return to that era. Nor can one ignore the terrible human and social toll that distress migration takes in India today — much of it for lack of water. "Dam it all" may not be a prudent course of action and is not anybody's case. But "damn it all" is a counsel of despair that forfeits the future which belongs to those yet unborn.

"Stop the Dam"

The survey of Bhakra-Pong, the Indira Gandhi Nahar, the Tehri Dam and the Narmada Valley projects contained in this volume does not suggest that these are either failures or wicked, dangerous, impossible, unviable or unsustainable. The critics and some of their international associates claim otherwise but have not been able to prove their case by rational argument. Some of their cautions, as on seismicity, are certainly valid but there is no reason to suppose that these were never considered or will be ignored. For the rest, the sweeping indictment of these projects is not well founded in fact. It rests on misconceptions, emotion, the magnification of initial delays or isolated events, confusing transient problems with permanent outcomes, and extrapolation of the past into the future as though there has been no learning experience, no technological development, and no wider awareness that will ensure greater accountability.

The critics' approach, barring the case of the IG Nahar, has not been problem-solving as much as coming up with a series of objections, contradictions notwithstanding, to damn the projects and "stop the dam". The attack on Bhakra-Pong is intended to disprove the success of the complex as this negates the arguments advanced against the other projects. The effort to debunk these large dams drives the critics to gross rhetorical excesses.

At one level it is sometimes suggested that the Tehri, Narmada and IG Nahar projects are perhaps not even necessary. Since this is difficult to sustain, alternatives are suggested. Not only have these sometimes been considered in the very process of project preparation — such as lower dam heights or other locations — but they are demonstrably less attractive, even if workable, and would entail far larger costs, displacement or submergence (in the case of small dams) for smaller benefits (as in the case of run-of-the-river hydel options) and would fail in critical periods when most in need. This is true of the many small dam and local water harvesting schemes advocated as alternatives to the SSP in Kachchh and Saurashtra. Not that these have not been tried and should not be further pursued where feasible. But these can only supplement and complement, not substitute, the Narmada Projects. Fallacies and myths should not be allowed to displace reason.

Of course there are costs in building large dams: human, ecological and financial. But these can be more than compensated. Those displaced or affected by projects, directly and indirectly, are now being offered generous R & R packages which have seen steady improvement and more rigorous implementation over time. There is little doubt that their condition will in due course be distinctly superior in relation to the living standards and opportunities they enjoyed earlier. This is already to be seen despite certain hard cases or problematic situations in the early phases which are amenable to correction.

Likewise, the ecological impacts are greatly outweighed by project benefits in the absence of which environmental degradation, migration and distress would take a further toll. The idea that events are standing still and that delays occasioned by further "reviews" or yet more surveys and studies do not matter and will improve the project is quite mistaken. Reviews and studies to the extent necessary can proceed alongside construction and project implementation except if critical to on-going phases of implementation. The *pari passu* principle adopted in the case of the SSP is perfectly valid and has prejudiced nothing that any critic has tangibly demonstrated up to date. On the other hand, every delay, apart from adding to costs, entails further prolongation of human misery and more environmental degradation and retards national development.

The benefits in each case are far greater than the costs, howsoever computed, and if social and indirect costs and benefits are taken into account with and without the project, the net gains would be all the greater.

It is in this regard that the earlier role of MOEF was not merely negative but inimical, especially in respect of the Tehri and Narmada projects. By divorcing environmental considerations from development it distanced itself from ground realities, emerging as the Abominable No-man so loved by some. Its stance was self-defeating and is now fortunately undergoing correction. Maybe all this manifests no more than the swing of the pendulum. The country and the world are hopefully moving towards a better balance on environmental concerns.

With the focus on proclaimed irrigation and power benefits, less notice has been taken of the seminal importance of providing drinking water supplies to large numbers through each of these projects, and more epecially the Indira Gandhi Nahar and SSP. Without water, millions would be on the move and on the streets demanding the most fundamental element of the right to life. The transforming nature of the development trigger each of these projects will release has not been understood despite the example of Bhakra-Pong. The value of peaking power and the role of hydro-electricity in stabilising the grid, upgrading the quality of energy supply and enhancing the profitability of the system is not appreciated. This does not mean that non-conventional sources of energy or gas should not be tapped or that energy conservation is unimportant. This is nobody's argument. There is need for all of these.

Wrong Calculus, False Promise

The positive impact of these projects on the environment and micro-climate and in creating oxygen banks and carbon sinks is scarcely realised. The Indira Gandhi Nahar and Sardar Sarovar will probably produce more greenery than lost to India since Independence as a result of all the other major water resource projects taken together. For every tree going under the SSP, about 100 will be newly grown. Today, water resource development is becoming a primary engine for greening India both in catchments as well as in commands.

The employment and income-generation potential of these projects is also surprisingly unseen. Irrigation augments the labour intensity of agriculture and, combined with flood control, groundwater recharge and power generation, creates a chain of jobs. One has only to wish away Bhakra-Pong as if a dream to realise with a shudder what the alternative would have been.

It is true that there are, and maybe will always remain, a gap between regions and generations over space and time. Hitherto, catchments have borne the major if not the entire burden of dam construction while the benefits have largely gone downstream or to downstream people who have usurped catchment area opportunities as well. Likewise, costs are borne today (including the emotional shock that comes with the first knowledge of possible displacement) whereas benefits may follow a decade or a generation hence.

The latter point is easily answered. If proper rehabilitation is ensured and people are well informed from the start about their likely future, they would by and large be prepared to make a generational sacrifice for the sake of their children even if they do not live to garner the full benefit of their contribution. As far as the first point is concerned, the upper catchment does benefit from catchment area treatment, afforestation, pisciculture and tourism and does generate some employment and dams often do lead to creation of infrastructure and area development of hitherto sequestered or neglected regions. But this is not nearly enough. It should be possible to impose a cess on power benefits or a betterment charge related to the appreciation of land values in the command, and make these sums available for overall development of the catchment. This could forge a new relationship between people above and below the dam and remove some of the resentments that obtain today.

Those displaced are being offered land for land in SSP, NSP and Tehri, though Tehri also has a suburban package near Dehra Dun which is proving increasingly popular among those who want to move off the land to non-farming occupations and self-employment. The R & R package provided to Tehri oustees has improved their living standards, as an independent evaluation has certified. In the case of SSP and NSP the package is even more generous in relation to land for land for landowners, major sons, encroachers landless and co-parceners in varying measure. Indeed, many in other projects elsewhere are demanding a similar package. Were there an abundance of land this would be ideal. But as things are, making land for land the primary if not the sole means of rehabilitation for all land-related categories is mistaken.

Where is the land? The land-man ratio has deteriorated sharply and there are increasing numbers of small and marginal farmers and landless in commands as much as in catchments. How are they all to be provided land, even if land ceilings can be effectively applied to newly irrigated areas? More so when under the SSP package the quantum of land required (quite apart from its quality) is likely to be far greater because of the wide eligibility criterion. There may be special reasons why it has been possible to develop a thriving land market in Gujarat where, as in the Tehri example, there are farming families which want to set up business or seek opportunities in town. But is this going to be replicable in areas with high population densities and fragmented farms?

Rather than offer false promises of land for land, the primary focus of R & R in the future must be employment, with land being one among several options where circumstances permit. The argument that traditional peasant and tribal communities are tied to the land and would be crushed if moved off it can no longer be sustained. If not out of choice then certainly of necessity, landless and marginal peasants and tribals who eke out a miserable living from unirrigated holdings, as in part of the SSP catchment, are migrating in larger and larger numbers, first seasonally and later for good as their traditional lands and occupations can no longer sustain them for even six to eight months in the year. If the rains fail, the crisis deepens. The Bhils in the SSP catchment in M.P. illicitly fell the forest in neighbouring Maharashtra in order to sustain themselves. In the Maharashtra submergence area new or expanding families merely clear more forest for cultivation. The forests are fast vanishing, and have totally disappeared in certain areas. As field and forest no longer serve, tribals by the thousands are moving out in distress as ecological refugees to towns and construction sites in M.P. and Gujarat where they face appalling problems of exploitation, cultural adjustment and social abuse. It comes then as something of a surprise that activist groups should spurn and thwart the far more benign option of orderly and gainful relocation of SSP oustees (readily admitting the real trauma of displacement), while the less informed IRM, western NGOs and the Washington-based Asia Watch should propagate this as forced eviction and a human rights violation!

Evoking Tribal Nostalgia

The attempt, particularly by the IRM, to divide Indian from (tribal) Indian and build up a case leading towards rights of "self-determination" and almost another sovereignty is mischievous and thoroughly misleading. There is no analogy between India's tribes (whose very rapid population increase itself sends out a positive message) and the "indigenous people" of the Americas and elsewhere who were systematically decimated, driven off the best lands and left to cope as best they could. It is only recently that a bruised social conscience has resulted in efforts to make amends while safeguarding the white outsiders' essential interests.

Tribal India is certainly part of have-not India. However, those exploited are not only tribals but other weaker sections of the population. Nor are tribal and non-tribal two ironclad categories. Acculturation and assimilation through sanskritisation has been evident over centuries. A lot remains to be done for tribal uplift. But glorification of their isolation and celebration of their recent graduation from hunter-gathering to simple agriculture will not do. They too must have access to education, better health and economic and social opportunity. The choice must be theirs. It is not for activist groups or the IRM to choose for them.

A sociologist close to the NBA, who has worked among the tribes being displaced by the SSP, brings out the underlying reality. "Why is the 'tribal' invoked if, and the Narmada Bachao Andolan acknowledges this, their reality is so heterogenous and their views of the world shot through with mediations by activists", asks Amita Bavaskar? "What purpose does the image serve? I think the Andolan used the 'tribal' as a symbol certain to appeal to our romantic notions, to fulfil our desire to hear something approximating Chief Seattle's speech. In a strange way, the 'tribal' has come to represent our normative self-image. This myth has been useful in the construction of an analytical framework by which to find answers to the 'development' dilemma. Thus the 'tribal' has been a cultural resource from which we have picked aspects that seem to fit our patchwork ideology . . . Reverence for nature, a relationship of intimacy with nature, a conception of harmony and sustainable resource use — all these are values that we appreciate in this post-modern world. That explains the resonance of the image/symbol". (Bavaskar, May 1991) How true.

There is no question that the weakest and poorest among those likely to be affected by dams or development merit the greatest consideration. Promises made on paper must be implemented with care and compassion and in good time. Initial delay in the case of Tehri and inter-state issues in respect of Pong and the SSP created problems not all of which have been fully overcome. But by and large R & R under these projects are now well under way. In the case of the Indira Gandhi Nahar, the problem is not one of rehabilitation (the Pong dam oustees apart) but of colonisation, as the desert must be settled if the newly opened IGN command is to be cultivated. While equity must always remain a prime consideration, efficiency cannot be ignored if the project is to succeed and equity realised. Putting equity before efficiency in the IGNP by settling landless farmers and cultivators lacking the experience and wherewithal to reclaim the undulating dune country and bring it under the plough, has not produced the best results. A mixed policy such as now being implemented through partial auctions is likely to prove more effective in expediting utilisation of irrigation, inducing more rapid settlement and generating funds for sooner completion of the project. This is a project above all others that requires the pioneering spirit.

Another Development Pathway?

"Stop-the-dam" critics have not hidden their true intent. In the case of Tehri there was a major point of difference regarding seismicity and dam safety. Otherwise, while individual aspects of these projects may have been the focus of dispute at one point of time or another, opposition essentially is to the projects per se as symbols of the current path of development which is felt to be wrong. The real battle is therefore about something very different. The object is not to improve these projects but to stop them. And since delay is a form of denial, the sniping on one count or another is part of a strategy of attrition which may be quite removed from fact or merit or any real effort at problem-solving.

Constructive critics like ARCH-Vahini at SSP and URMUL at IGNP have on the other hand supported the projects when genuine concerns have been met.

Since the pattern of development is what is essentially being questioned, this argument should be examined. The NBA states without any ambiguity that the Andolan "has emerged as a major challenge to the prevalent notion of development". The opposition is to a "centralised, capital-intensive, consumerist, urban-oriented policy sustained by a continuous drain of resources from rural and tribal hinterlands with the connivance of multilateral agencies". The target is the "fossilised, outdated, Western concept of Development and technological progress, where growth is God". (NBA, 1991-92) "A tomb of destruction: based on an obsolete model of development being propelled by a deplorable combine of a section of corrupt politicians, bureaucrats and greedy contractors" at the cost of the people, is how Baba Amte describes the SSP. (July 1993) Elsewhere, he calls big dams "monstrous cruelties" and asks why the tribals who are being displaced by the SSP should leave "an entire civilisation. . . . because four foolish chief ministers have signed some accord". (August 1993) Sunderlal Bahuguna inveighs against "economic growth. . . the new religion of humankind with the dollar as the new god". He would leave the Himalaya as "a place where people from the disturbed world would go in search of peace". (July 1993) For B.D. Sharma, "Harsud (in the NSP command) has become a symbol of national struggle, particularly of the tribal people against the present style of development . . . (with) equity relegated to a secondary position". (August 1991) And, "The Narmada project is part and parcel of India's post-Independence elite-oriented development path". (Alvares and Billorey, 1988)

Sample some statements by Medha Patkar. Attacking the Calcutta Electric Supply Corporation's Budge Budge thermal project, currently under way, she said that protecting the ownership of the land "rather than providing jobs" was the road to security. (October 1993) The white revolution has reduced milk intake in certain areas "because you compel . . . (the farmer) to fall a prey to (the) market economy. It is a question of what model of development you choose". (November 1992) "The agitation has the larger goal of demanding development strategies that are not just eco-friendly but are actually man-friendly. We want the entire process of so-called development, which has proved disastrous for the people, to change". Medha Patkar, like Maneka Gandhi, is for developing the Green movement into a political force. (Patkar, July 1993; Gandhi, Maneka; August 1990)

Finally, an American NGO view: "The Narmada movement evokes many images: the dams as avatars of destructive development" with "local populations long relegated to the margins of power and history, mobilising to defend the ecological balance of the resources on which their survival depends" against the threat of "a globally hegemonic economic system that is colonising and dominating the last bio-geographical and cultural spaces of the world in the name of development". (Rich, Bruce, May 1991)

Answering the Critics

This random sampling makes interesting reading. What does it say? With some statements there can be no disagreement in principle. Thus a Green movement, even if in politics, would be welcome. Development should certainly be man-friendly. Indeed, uplifting the underprivileged and imparting a degree of equity to the economic process is implicit in development as opposed to "growth". This has, alas, not been achieved and continuous striving in that direction is a widely-shared objective. Development must be eco-friendly and sustainable too. That again is now more widely understood and accepted. But are Tehri, the IGNP or the Narmada projects not man or eco-friendly? The notion that they largely serve the rich and exploit the marginal and poor is a travesty. The belief that a market economy is per se bad and a pre-modern subsistence economy or primitive exchange based on self-reliance and barter is somehow nobler, is scarcely to be taken seriously. The reason that Operation Flood farmers sell milk rather than drink all of it, is because they today produce more than they can consume and market some or all of it in order to satisfy other needs. They are better rather than worse off for the transaction by any reckoning and they enter the market not out of necessity but out of choice. And why is employment (jobs) inferior to a small or marginal land holding? There is not enough land for everybody and fragmentation and dimunition of holdings is rendering farming uneconomic. It makes good sense to take people off the land and provide them income and security through other forms of gainful employment.

Yes, small may be beautiful. But does this make big bad? There is no way Gujarat can harvest 9 MAF of Narmada water (or secure this from any other untapped source within the state) without the SSP. And to think exclusively of substitution when the need is for substantial additionality is to forego both eco-friendly and man-friendly development. It is not appreciated that the Narmada Valley master plan envisages 30 big dams, 135 medium dams, 3000 small dams and maybe 30,000 micro water-harvesting or conservation schemes. This is massive small scale development. And even the SSP command is proposed to be managed by water user associations of farmers buying water wholesale at lower offtake points and thereafter managing retail distribution and drainage on their own within small farming communities. Tehri will energise thousands of tubewells and agro-processing and rural industries, as does Bhakra-Pong power.

Why should Baba Amte be oppressed by the "monstrous cruelty" of large dams when millions regard them as a blessing? He made a brave effort earlier to bring peace to Punjab when it was in turmoil. Did he then think that the cause of peace would have been advanced by dismantling Bhakra? The four "foolish" chief ministers whom he chides for resolving the Narmada dispute were indeed "foolish" not to have come to such an understanding 20 years earlier, in which case SSP and NSP might have been fully operational today.

Sunderlal Bahuguna would preserve the peace of the Himalaya. Let him ponder the facts. "There are instances in the Garhwal Himalaya where more than three-fourth of the families migrated out of the village and an equal proportion of cropped land was abandoned during the period 1972–88. It is for salaried jobs (that Medha Patkar so despises in favour of land ownership) for which males migrate, but ownership of native land is retained. Management of land in such situations becomes the exclusive responsibility of women. Enormous expenditure of energy and time in collecting fuelwood, fodder and drinking water have made life miserable, specifically of rural women". (G.B. Pant Institute of Himalayan Environment and Development, 1992) Greater gender equity, relief from grinding drudgery and a real chance for the girl-child are likely to be major benefits from these projects in Garhwal, Rajasthan and Saurashtra-Kachchh.

Those who would romanticise the tribal condition are very distant from reality. Eco-environmental studies of the SSP catchment yielded this analysis: "It is not possible to accept the concept of environment versus development. . . . For successful defence of the environment in rural areas it is necessary to develop and organise them to a point where they can provide services and goods to other areas and in their turn receive those goods and services that they need. In other words, it is the rural isolation and their pathetic attempts to be self-sufficient even at sub-subsistence levels, while providing for outside systems, that is responsible for environmental degradation and comes in the way of rebuilding the eco-system (*sic*). (MSU, Baroda, 1992)

Some of the critics would benefit from listening to the Norwegian Prime Minister, Gro Harlem Brundtland who headed the World Environment Commission which preached sustainability in its Report, "Our Common Future": "Pro-conservation movements must not become so over-eager that they turn against poeple. Many, including indigenous people, have felt the protective decisions taken far away, threaten their very existence". (July 1993) She was writing about the protection of endangered species. But the caution has wider application.

There is a cost for doing anything. It is, however, erroneous to imagine that inaction entails no cost. On the contrary, doing nothing can be far more expensive and injurious. It is sometimes questioned whether India can afford these large projects. The burdens are obvious and many sacrifices will have to be made. But not merely will the gains be greater by far, but it might more appropriately be asked if India can afford not to build these projects? Gujarat spent over Rs 1500 crores in direct official relief coping with three years of drought during the period 1985-87 while crop and other private losses are estimated to have been of the order of Rs 5000 crores. Is this kind of expenditure affordable or avoidable?

Global Experience

The World Bank in a new policy paper on water has estimated a requirement of $ 700 billion to be invested over the next decade to ensure inexpensive, clean and reliable water supplies for the diverse needs of peoples in the developing world. If world population increases to 8 billion by 2025, as expected, with almost a seventh of that in India, the demand for water for irrigation and domestic and industrial use will grow by 650 per cent over the next 30 years. Urbanisation will exacerbate the problem. (*Hindustan Times*, September 1993) India is already facing a water crisis and delay in coming to grips with the problem can only lead to increasing social unrest, adverse health impacts, economic loss and violence. By all means let all sources and technologies, big and small, known and non-conventional, be exploited. But to exclude an obvious, cheap and most effective method from the start appears suicidal — a *jal samarpan* in reverse. There is danger in global "water blindness" — failure to see a crisis that stares one in the face. (Clarke, Robin, 1991)

Foreign experience is instructive. The Aswan High Dam was widely cited as an environmental disaster three decades ago when it was mooted and under construction. All manner of dreadful consequences were forecast that spelt the doom of Egypt. What has been the record? A quarter century after the dam has been put to work the Executive Director of UNEP had this to say: "There is no argument: the dam's contributions to substantially increasing food production and hydropower generation has made an enormous difference to the quality of life of the Egyptian people . . . And in this particular case the adverse impacts were all anticipated and plans set for addressing them, but international political situations delayed the implementation". He urged that positive impacts be seen alongside negative impacts and balanced decisions taken. (Tolba, 1990) Another opinion: "The evaluation of the past 21 years since the completion of the high dam at Aswan has found that the project has had a profound impact on Egypt and the Nile. Irrigation, power and flood control benefits have all been achieved and are large. The dam gave important protection to Egypt during the African drought and, in particular, during the critically "dry" year of 1983. There are some who believe that the dam paid for itself in that one year that was almost as bad as 1913". (Dixon, Talbot, Le Moigne, 1990) An assessment made of the Aswan High Dam by a joint American-Egyptian team over seven years looked at the positive and negative impacts of the project. The conclusion was unambiguous: "the Aswan High Dam is not an ecological disaster but an economic blessing". The Aswan hydro-power station was found to be the most profitable capital investment in national history, having paid off in five years while saving the Valley from floods and feeding Egypt. (Bauingenieur, 1982)

These comments are interesting as the Aswan Dam, like the Narmada Valley projects today, was pilloried by critics who would have stopped it if they could. The Doomsday people have been proved resoundingly wrong. But the same dire forecasts are being made about Narmada. The Clinton Administration was reported planning to tell the World Bank to re-examine its funding of large scale power projects in developing countries "and work energetically to avoid the possibility of more Narmadas in the future". (Bhushan,1993) World Bank projects, it was said, must advance human rights and not harm the environment. The advice, if actually given, was based on a completely wrong appreciation of both facts and the needs and shows how many have fallen prey to the flawed thesis of the IRM.

Western governments, sensitive to Green pressures, have been strong on cautioning against dam construction in the Third World having put the bulk of their water resource development behind them. But when they see economic opportunity in their own country or from possible overseas investments, how do they react? In November 1990 the Royal Canadian Mounted Police was reportedly employed in the Peigan reserve in Alberta to remove indigenous communities agitating against the Oldman dam which threatened forest submergence. President George Bush in 1990 put jobs and the well-being of American families ahead of the spotted owl which environmentalists wanted declared an endangered species and preserved through a ban on logging in Oregon and Northern California. In early 1991, *The Economist* of London reported that the director-general of the Japanese environmental agency lost his job because he publicly opposed a dam across the Nagara river, (the last of Japan's 139 major rivers to be harnessed). Japan had just suspended assistance to the Narmada power plant on environmental grounds. Logging continues on Fraser Island, off the coast of Western Australia, despite long-standing protests by local rain-forest groups. (Agarwal and Narain, 1992) German industry, which is being wooed to invest in India, recently proposed that the Indian government should give automatic environmental clearance if the investor is in possession of technological clearance in Germany. (*Business and Political Observer*, September 1993)

A Vision of the Future

Every nation needs a vision of the future. This cannot be exploitative or self-destructive. Equity and sustainability must be prime concerns. Yet, the future cannot be based on nostalgia for a distant or frozen past. Disenchantment with the existing order and current development models can be no reason to cling to a past that cannot be recalled because social and technological change, the exponential growth of numbers, globalisation and security considerations render hunter-gathering or subsistence economies totally obsolete. Nor need one fear innovation and technology and behave like latter-day Luddites who would turn their backs on the future. One can learn from the past, not return to it. The large water resource projects that have come under ferocious attack are not monumental follies. They offer hope to millions. They are not examples of consumerist greed in the current Indian context.

The divisibility and reach of technology and globalisation and the transnational corporation have between them begun to challenge the concept of the nation state. Larger political, security and economic structures have evolved, of which the European Community and ASEAN are examples. Empires have crumbled, the Soviet Union being the latest, while Yugoslavia has fallen apart. In the groping towards a new order, concern for the identity of small groups and communities is manifest. This global reversion to tribalism finds an echo in nostalgia for (and, for some, guilt about) a tribal past. There is disenchantment with the present order, apprehension about gains from post-modernism and globalisation for the few at the cost of economic servitude for the many. There is a search for new directions. The quest must go on. And India will make its own explorations, possibly along the line of decentralising several functions through panchayati raj and community initiatives. But the "people's survival strategies (which) reverse the overwhelming and encroaching power of the world state system", as advocated in the Bangkok Declaration of the People's Forum which met in October 1991 on the occasion of the annual Fund-Bank meeting, scarcely sounds an effective alternative except for small and relatively simple communities. Not that this offers a panacea either. Life styles must change, with more for the truly needy and less of ostentatious consumption in profligate societies or Third World enclaves. Gandhi's dictum that there is enough for everybody's need but not for everybody's greed is eternal and universal.

This does not, however, mean renunciation or acceptance of austerities and deprivation that derogate from a fair quality of life. Nor does it mean abandonment of efforts optimally to harness the country's water and other natural resources to achieve this end. There is a right to development for all with dignity and equity. Despite its considerable resources and truly rich human talent, India remains the largest single reservoir of poverty, illiteracy and disease in the world. In relative terms, it has regressed while nations far behind it have moved ahead. This can scarcely be a matter of pride. The answer therefore lies in doing more and better without delay rather than otherwise. India has to be bold and innovative, willing and able to reach out and grasp opportunity and not shrink, and be afraid to do. It has to be a confident, achieving and yet ecologically responsible society. Narmada, Tehri and the IGNP, like Bhakra-Pong, are milestones on the way.

REFERENCES

1. Wittfogel, Karl A., The Hydraulic Civilisation. From *Man's Role in Changing the Face of the Earth*. Edited by William L. Thomas Jr. University of Chicago Press, 1956.

2. Narmada Bachao Andolan, "Towards Sustainable and Just Development, The People's Struggle in the Narmada Valley". Dhule, 1991-92.

3. Amte, Baba, Amte Accuses Centre of Indifference. U.N.I. *Business and Political Observer*, New Delhi, July 28, 1993.

4. Amte, Baba. I see an Uncommon Strength in the Common Man. Interview with Dibang in the *Illustrated Weekly of India*, Bombay, August 7, 1993.

5. Bahuguna, Sunderlal, Damning Blunders: Himalayan Ecology. *Hindustan Times*, Delhi, July 9, 1993.

6. Rich, Bruce. Narmada and the Global Battle for the Biosphere.

Environmental Defence Fund, Washington. Article in *Lokayan Bulletin*, New Delhi, May-August 1991.

7. Patkar, Medha. Medha Warns WB Power Unit. *The Pioneer*, New Delhi, October 3, 1993.

8. Patkar, Medha. Cited in "Narmada Dam: Economic Boom or Ecological Disaster." Crossfire column in *India Today*, New Delhi, November 15, 1992.

9. Patkar, Medha. Medha cautious, but hopeful after talks. *Times of India*, New Delhi, July 5, 1993.

10. Patkar, Medha. Green movement will turn political. *Times of India*, July 9, 1993.

11. Gandhi, Maneka. Green must replace greed. Interview with N.D. Nalapat, *Times of India*, August 12, 1990.

12. G.B. Pant Institute of Himalayan Environment and Development. Action Plan for Himalaya. Almora, 1992.

13. M.S. University. Eco-Environmental Studies of Sardar Sarovar Environs. Project Director, S.D. Sabnis, Department of Botony and Zoology, Baroda, 1992.

14. Brundtland, Gro Harlem. Pro-conservation movements must not turn anti-people. *The Pioneer*, New Delhi, July 8, 1993.

15. Clarke, Robin. *Water, The International Crisis*. Earthscan Publications, London, in collaboration with the Swedish Red Cross, 1991.

16. Tolba, Mustafa Kamal. 'Foreword' to *Environmentally-Sound Water Management* by N.C. Thanh and Asit K. Biswas, Oxford University Press, Delhi, 1990.

17. Dixon, John A., Talbot, Lee M., Le Moigne, Guy J.M. *Dams and the Environment*. Considerations in World Bank Projects. Washington, 1990.

18. *Bauingenieur* Magazine, Published in Germany, Issue No. 12, 1982.

19. *Hindustan Times*. World Bank announces new water policy. N.C. Menon, Washington, September 17, 1993.

20. Bhushan, Bharat, No more Narmadas, Clinton Administration tells World Bank. *Indian Express*, New Delhi, 1992.

21. *Business and Political Observer*. Kohl asks Rao to raise royalty ceiling, by Sukumar Sah. New Delhi, September 14, 1993.

22. Agarwal, Anil and Narain, Sunita. The West should look at its own backyard first. Green Politics. *Economic Times*, New Delhi, May 1992.

23. Langan Fred. Canada gives home rule to its Eskimos. *The Pioneer*, New Delhi, June 7, 1992.

24. Seth, S.P. Australian Aborigines wait for justice. *The Pioneer*, Delhi, July 22, 1993.

25. Postel, Sandra. *Lost Oasis, Facing Water Scarcity*. The Worldwatch

Environmental Alert Series. Horizon India Books, New Delhi, 1992.

26. *Times of India* News Service. "Drought spells cause havoc in Saurashtra". Jamnagar, October 25. *The Times of India*, New Delhi, October 26, 1993.

27. *The Pioneer*. Water Crisis: Poll Issue and Perspectives. New Delhi, October 22, 1993.

Postscript

CERTAIN developments since this volume went to press are encapsulated in this summary update.

In its search for additional funding for the Indira Gandhi Nahar Project, the Rajasthan Government has sought to interest the World Bank in the scheme. WAPCOS had at the Bank's instance done a Revised Feasibility Study for IGNP Stage II which at 1991 prices had been costed at Rs. 3549 crores, Rs. 1990 crores for the canal system and Rs. 1559 for command area development. At the present rate of progress, constrained by funds, WAPCOS envisages a completion date of 2006-07. Accounting for cost escalations it has estimated the final cost of Stage II at Rs. 6247 crores.

The IGPN command is saucer shaped and has no natural drainage outfall. This has resulted in waterlogging and salinity in certain parts of Stage I and II as earlier described. The slope of the land is southwest, but Pakistan has legitimately constructed a bandh along its border.

Draining the IGNP Saucer

The Canadian International Development Agency has from 1991 been associated with a programme in the Chambal command for the planning, design and construction of a sub-surface drainage system over 25,000 ha to remove salinity. One component of this is to look at the feasibility of a similar solution for portions of the IGNP command. Since the World Bank is funding a larger Agricultural Development Project which takes in the IGNP, the two agencies have come together to undertake this particular $ 3 million study.

The Bank's terms of reference include four main areas of concern, namely, the drainage of IGNP command, a "sink-hole" into which over 6 MAF of water is being poured annually; the extremely poor soils and dune-covered nature of the command, especially in Stage II, which pose problems of non-irrigability by traditional surface methods and of maintaining open channels liable to choke with wind-blown sand; the delicate ecological balance of grassland plants in Stage II which stabilise the soil but must be cleared for cultivation; and the need for a comprehensive environmental impact assessment giving rise to a rational plan for sustainable agriculture before further development is undertaken.

The IGNP authorities, however, contend that whereas an earlier FAO report had concluded that large tracts of Stage II were unsuited to traditional surface irrigation, its own more recent soil surveys over 300,000 ha give a contrary finding. The proposed Bank-CIDA study, for which consultancy bids have been invited, may therefore proceed parallel with the opening of new areas to irrigation as various Stage II distributaries are ready for commissioning. The Bank's study includes a pilot sub-surface drainage project in a comparable 5000 ha of Stage I, the efficacy of evaporation ponds and other measures.

Vertical drainage by itself may not work as mineralised drainage flows, even if treated and recycled, must finally find safe disposal. Alternative cropping patterns such as pasture and horticulture with sprinkler and drip irrigation could provide an answer. Israeli expertise might be summoned. The CWC has suggested laying underground pipes for lower distributaries in order to reduce seepage and overcome maintenance problems.

The IGNP is not unaware of these problems and possible solutions and has already moved to reduce the water allowance in Stage I and II. However, there is no room for complacency and all the questions raised must undoubtedly be addressed to safeguard the future of the project without diminution of its great benefits.

The compulsion to hurry and use all the available water, as explained, stems from the apprehension that, given the hard-won water allocation deriving from the Indus Water Treaty, failure by Rajasthan fully to utilise its share could create user rights in Punjab or Pakistan. This may be an exaggerated fear. If the drainage problem is insuperable or if effective solutions are delayed, other alternatives suggest themselves. Rajasthan could trade a million or more MAF for some time, or permanently, with Punjab-Haryana for additional power generation and a larger share of electricity or other benefits. Or, it could through the Government of India negotiate a water-for-drainage swap with Pakistan by a suitable amendment of the Indus Treaty. Pakistan has similar salinity problems in the lower Indus basin and has sought to solve them by constructing a drainage outlet from Nawabshah to the sea. It is water short and should be glad to receive more water for its use down the Sutlej in lieu of which it could give Rajasthan a drainage outlet to its Nawabshah outfall or otherwise. The trade-off would be mutually beneficial. Drainage of saline water from the IGNP command to the Rann of Kutch could be far more problematic and would entail some lift.

Tehri Collaboration

The Tehri project too has yet to overcome funding problems. The indigenously-financed component is going ahead. The coffer dam should be ready by June 1994 and R & R of the township and coffer dam PAPs continues. The Phase-I power house component may now be taken up with a Russian-Ukranian $ 100 m package on deferred terms. On catchment area treatment, MOEF has apparently finally accepted that only those areas found to be critically degraded through remote sensing need to be treated, as argued by THDC.

Agitation against the dam continues fitfully not on the seismic issue, which hopefully has been finally laid to rest, but on the plea that the R & R terms must be upgraded to match the Narmada (SSP) package. This is not a serious proposition.

First SSP Benefits in Sight

This finally leaves the Narmada Valley projects. The Environmental Sub-Group of the Narmada Control Authority was at the turn of the year reported to have favoured a halt to construction of the Sardar Sarovar dam as MOEF felt that R & R and environmental impact studies had failed to proceed pari passu with the raising of the dam. Plugging the 10 construction sluices in the apron, 18 metres above the river bed was feared likely to have adverse effects on villages that had not yet been rehabilitated as impounded water would rise to the river sluice level at 53 metres.

These matters were, however, quickly resolved at a special high level meeting taken by the Prime Minister on January 12,

1994. MOEF was directed to release a further 1500 ha of degraded forest land at Taloda in Akkalkuwa taluka in Maharashtra for R & R of sluice-aftected and other persons. The condition imposed is that GOM must allocate appropriate lands for compensatory afforestation and that the revenue derived from clear-falling of the additional Taloda forest now released must be deposited with MOEF. All other environmental parameters will, as before, continue to be strictly monitored so as to ensure compliance with the agreed norms.

Earlier, there had been renewed NBA-led agitation in Dhule over the survey of villages likely to be temporarily or permanently affected by closure of the construction sluices which are no longer required and were scheduled to be plugged in order to permit completion of the lower spillway of the dam before pitting and scouring occurred in that section with the raising of the dam. Survey teams were prevented entry into the villages by stoning, causing injuries and compelling resort to police firing resulting in the unfortunate death of a youth. The NBA immediately staged protests and demanded an independent inquiry but the Maharashtra government said that the incident had been video-taped and police action followed clear provocation.

The agitation against plugging of the construction sluices was again based on misperceptions. Far from swamping a large number of villagers whom it would not be possible to rehabilitate before the 1994 monsoon, the problem was more manageable. The construction sluices can only pass 30,000 cusecs of water and their plugging will therefore raise the flood level by no more than 30 cm. So no great danger portends. With the promised release of an additional 1500 ha of Taloda forest, it will be possible to provide permanent rehabilitation to those families whose habitations are likely to be permanently submerged in 1994 and temporary R & R to those who might face temporary submergence on account of the backwater effect. The lowest part of the SSP dam at the end of 1993 was at 69 m and R & R had already been completed up to that contour level, with a little safety margin. With the release of the additional Taloda forest land, the minimum crest level of the dam can safely be raised to 87 m before the ensuing flood season.

There is no substance in the charge that catchment area treatment has not kept pace with the raising of the dam. This has been done pari passu with construction in relation to the directly draining area of the catchment as recommended by the Dewan Committee. MOEF, however, desires CAT to be completed within the entire free draining area which would be impossibly large and was in fact ruled out years ago.

The other point made out was that there could be serious downstream effects if the construction sluices were plugged, especially with the drying up of the river for several weeks which would affect drinking water supplies and kill all fish. Downstream flows would indeed be interrupted between the plugging of the sluices at 18 m and filling of the reservoir to 53 m when water would once again flow through the river sluices. But this would only be for 10-12 days and not only had the Gujarat government positioned equipment to pump sufficient water over the dam for this period, but the nearby Karjan reservoir on a tributary stream lower down had been filled to augment the lower Narmada. As usual, NBA spokespersons made apocalyptic statements but the State and Central Governments stood firm.

Work on the Project continues apace. The good news is that the SSP Nahar Nigam expects to complete all works in time to commission Stage I irrigation up to the Mahi River by June 1995. The Japanese Government has reportedly been approached to restore the balance of a $150 m funding for the canal power house, earlier suspended on account of environmental apprehensions. The turbines ordered through Sumimoto against that credit have been ready for some time and given early release of funds could be shipped and installed to begin generation within some months as all the civil works are complete. This could be an important landmark as power generation will earn instant revenue and assist in the further financing of the project as planned. In any event, even the partial commissioning of Stage I after the many vicissitudes faced by the Project will give all those behind the SSP a tremendous psychological boost.

Work has also commenced on the first phase of the project to supply drinking water to Saurashtra and Kutch from the SSP canal system. Rs. 100 crores have been budgeted for 1994-95 and tenders invited. First supplies are expected to be delivered within three years by pipeline. Here is another supremely important facet of the SSP project, termed "impossible", taking shape. This project is being undertaken by the Gujarat State Construction Corporation.

Funding, though tight, will be somewhat eased should Japanese aid be restored. The revised cost of the project at 1989-90 prices is now placed at a little over Rs. 9000 crores. The first Sardar Sarovar bond issue of Rs. 300 crores was oversubscribed to the extent of Rs. 274 crores and the Nigam has been authorised to retain up to 25 per cent of the excess. Thus far Rs. 332 crores have also been raised through bank credit, bridge finance loans, private placement of debentures, public deposits and inter-corporate deposits. Rs. 815 crores was due to Gujarat from Maharashtra, MP and Rajasthan as their share up to October 1993. Of this Rs. 490 crores had been received by Gujarat leaving a balance of Rs. 326 crores of which just under Rs. 70 crores was disputed. A sum of Rs. 2864 crores had been spent by the Nigam up to December 1993. Rs. 1200 crores is budgeted to be spent during 1993-94 and another Rs. 1600 crores in 1994-95.

Meanwhile, the Narmada Eminent Persons Group set up in the summer of 1993 has yet to submit its report which has been delayed by a stay order in the Gujarat High Court. Whatever the findings, its labours appear to have been overtaken by events, something that seemed obvious at the outset when Medha Patkar kept insisting that it was a "review committee" while the Government firmly told Parliament it was not. The Group can at best advise on future environmental and R & R aspects in the light of its examination of the record up to date. It may also suggest guidelines for the future which would be useful.

NSP Progress and GOMP Feelers

As far as the Narmada Sagar Projects in Madhya Pradesh are concerned, the ISP coffer dam and diversion tunnels have been completed along with the Maheshwar coffer dam. The Technical Committee of the CWC has cleared the Omkareshwar Project and environmental clearance is awaited. It has once again been mooted by GOMP that the height of the SSP might be lowered. It had earlier been argued that a height-reduction of 19 feet would safeguard all the projected irrigation in Gujarat and reduce power generation by only 14 per cent, of which MP's share is 57 per cent which it would be prepared to forgo. The NWDT had rejected this as sub-optimal.

It is unlikely that Gujarat will agree to a lowering of the height of the dam. Maharashtra too might hesitate to reverse gear and lose on power when its R & R goal is in sight with the release of additional Taloda forest land. If, however, a token height-reduction of four or five feet can save face all round and expedite progress, this might be a proposition that some might feel worth examination. For, beyond a minimum crest level of the order of 100 m, any further raising of the Sardar Sarovar dam to its designed height will have to move pari passu with R & R in Madhya Pradesh. Delay could slow down project completion though without affecting the early phases of SSP irrigation.

While Madhya Pradesh would save on R & R from the sometimes-projected lowering of the height of SSP, it will have to forgo the additional power gain as well as potential inland navigation benefits from a longer and deeper waterway that the designed dam height would ensure. Use of such a waterway is low in everybody's calculus just now, but the future benefit to M.P. and India from such a cheap transportation artery reaching deep into its heartland could be very significant. Any reduction in the height of the SSP would correspondingly reduce carryover storage, which would again be a very considerable loss in an emerging scenario of water shortages, especially in the case of a terminal dam. Finally, the consequent loss of power would materially affect the ultimate benefit-cost ratio of the project.

The same arguments regarding power benefits, carryover storage, navigation and economic viability would apply to any reduction in the height of the Indira Sagar Dam.

The new MP Chief Minister has entered a plea before the Tenth Finance Commission that the NSP be treated as a national project and be Centrally funded. This point has been made from time to time on behalf of other projects too, including the IGNP, but has little chance of being conceded at this juncture. More to the point, the pricing of power and water should be realistically reviewed so as to cover operational and maintenance costs and meet the interest liability on the capital invested at the very least. Unrequited subsidies, like other populist measures, are unsustainable and ruinous in the long run.

In a conciliatory move, the MP Government has invited Medha Patkar, the NBA leader, to join the resettlement monitoring group constituted for the Bargi Project. She could make a constructive contribution here.

Two Global Pointers

Even as the controversy about the Narmada Valley, Indira Gandhi Nahar and Tehri projects have raged, the Population Action International in a study entitled "Sustaining Water: Population and the Future of Renewable Water Supplies" released in Washington (*Hindustan Times*, November 7, 1993), has warned of a massive global water scarcity. It said that in 1990 more than 335 million people lived in water-stressed or water-scarce countries or regions. It projected an increase in this figure with population growth to between 2.8 and 3.3 billion people by 2025, including India. Those who denigrate large water resource projects might find it instructive to read this and similar reports. The "Sahelisation" of a growing number of regions could be an unmitigated human disaster. The Narmada Valley, Indira Gandhi Nahar and Himalayan storage projects like Tehri are designed to avert any such catastrophe in India. The gestation period for so doing being long, one must act now and urgently to safeguard the future.

Critics may also like to take note of the remarks made by J. Cotillon, Secretary-General of the International Commission on Large Dams at the 61st executive meeting of ICOLD in Cairo on November 3, 1993 at a symposium on the High Aswan Dam (HAD). He confessed to his own misconceptions, as of countless others worldwide, regarding the project and extolled its manifold successes of which he had now become aware. He praised the courage, devotion and faith in the future "of all the Egyptian engineers and people who made the dam and have had to put up with so much criticism." His conclusion: Now there can be "legitimate pride in this magnificent, necessary, vital and urgent engineering achievement . . . giving life, security and prosperity to Egypt's land and people. The maligned HAD has become one of the landmarks of Egypt's brilliant history." This is fulsome praise after many years of calumny, denigration and virulent opposition as an environmental disaster, a judgement now shown to have been monumentally wrong.

No great endeavour is without its challenges, trials and tribulations. Prudence and care are always necessary. A generation hence, properly managed, the blessings of these great new projects, Narmada Valley, Indira Gandhi Nahar and Tehri will, like Bhakra-Pong, be seen as vital life-support systems for a resurgent India and millenary gifts to the nation. The future beckons.

New Delhi
March 1, 1994

Index